Oscillators - Recent Developments

Edited by Patrice Salzenstein

Published in London, United Kingdom

IntechOpen

Supporting open minds since 2005

Oscillators - Recent Developments
http://dx.doi.org/10.5772/intechopen.75318
Edited by Patrice Salzenstein

Contributors
Roman Ivanovich Parovik, Coşkun Deniz, Ozkan Ozturk, Yonggang Tan, Vladimir Yurchenko, Lidiya Yurchenko

First published in London, United Kingdom, 2019 by IntechOpen
IntechOpen is the global imprint of INTECHOPEN LIMITED, registered in England and Wales, registration number: 11086078, The Shard, 25th floor, 32 London Bridge Street
London, SE19SG - United Kingdom
Printed in Croatia

British Library Cataloguing-in-Publication Data
A catalogue record for this book is available from the British Library

Additional hard and PDF copies can be obtained from orders@intechopen.com

Oscillators - Recent Developments
Edited by Patrice Salzenstein
p. cm.
Print ISBN 978-1-78985-837-2
Online ISBN 978-1-78985-838-9
eBook (PDF) ISBN 978-1-83881-068-9

Meet the editor

Patrice Salzenstein holds a Masters and engineering diploma (University of Lille, 1993) and a PhD degree in Electronics (University of Lille, 1996). Between 1996 and 2001, he worked near Paris successively as a technology research engineer at THOMSON-CSF-LCR (Thales-TRT), microwave engineer at ALCATEL, and head of the RF and Time & Frequency Group at LCIE private research laboratories. Since 2001, he has been working with oscillators for the Centre National de la Recherche Scientifique (CNRS) at FEMTO-ST institute in Besançon, France. In 2010, one of his articles was featured in Electronics Letters for his participation with Czech and Swiss colleagues regarding the best frequency stability ever measured on a quartz-crystal oscillator: 2.5×10-14 at 5MHz. His fields of interest in research were for optoelectronic resonators and oscillators (2010–2017) and now on Brillouin light-scattering and instrumentation as a CNRS senior research engineer. He is author or coauthor of more than 22 high-impact factors papers.

Contents

Preface

An oscillator is dedicated to the generation of signals. It is used in computers, telecoms, watchmaking, astronomy, and metrology. It can be a pendulum, an electronic oscillator based on quartz technology [1], an optoelectronic oscillator [2], or an atomic clock, depending on its application. Since water clocks of antiquity, mechanical clocks invented during the thirteenth century, and the discovery of piezoelectricity by Jacques and Pierre Curie in 1880 [3], oscillators have made great progress [4]. This book does not attempt to tell the story of oscillators, but rather to provide an overview of particular oscillator structures through examples from mathematics to oscillators, and from the millimeter scale to the vibration of a building, focusing on recent developments, as we live in a time when technology and mathematical analysis play a vital role.

By providing an overview of the quest of mathematics for the oscillator, we had to make choices, as is evident in the selection of the first three chapters:

- It all starts with the need to have the best mathematical models to understand what an oscillator is. This is why we became interested in the work of Roman Parovik, researcher at the Institute of Cosmophysical Research and Radio Wave Propagation at Vitus Bering Kamchatka State University, Petropavlovsk-Kamchatskiy, Russia. He investigates mathematical models of oscillators with memory.

- Then we focused our attention on the research of Coskun Deniz, working as Lecturer in Department of Electrical-Electronics Engineering in the Aydin Adnan Menderes University, Aydin, Turkey. He has worked on the quantum harmonic oscillator (QHO) and two conventional semi-classical approximation methods to solve QHO.

- Ozkan Ozturk from the American University of the Middle East in Kuwait is interested in theories of time-scale systems to acquire information about the long-time behavior of nonlinear systems. Therefore he is interested in the oscillation criteria of two-dimensional time-scale systems.

We have chosen to emphasize the place of oscillators in our word. The oscillation problem begins at the micrometer scale with microstrip oscillators and extends to the macroscopic scales with civil structures such as bridges. Two following chapters discuss these themes:

- Time-domain simulation of microstrip-connected solid-state oscillators for radar applications is a hot topic investigated by Vladimir Yurchenko and Lidiya Yurchenko at the O.Ya. Usikov Institute for Radiophysics and Electronics, Kharkov, Ukraine.

- Yonggang Tan, awarded by the Chinese construction Enterprise Management Association of Science and Technology, works on bridge and tunnel engineering at the faculty of infrastructure engineering, Dalian University of

Technology, Dalian, China. He is interested in reducing vibrations of civil structures under harmonic or wind excitations.

Authors and coauthors cover various areas of oscillator research throughout this book. The common thread of the chapters in this book is to conduct investigations in the field of oscillators. This book will certainly be useful for students, engineers, and researchers who want to keep up with the latest developments in oscillators through chosen examples. We assume that the original approach of this book consists in the possibility offered to the reader to realize the diversity of the field covered by oscillators and their relevance in research today.

Patrice Salzenstein
Senior Research Engineer,
Centre National de la Recherche Scientifique (CNRS),
Franche-Comté Electronique Mécanique Thermique et Optique—Sciences et Technologies (FEMTO-ST) institute, mixed research unit associated with CNRS (UMR 6174),
Besançon, France

References

[1] Salzenstein P, Kuna A, Sojdr L, Chauvin J. Significant step in ultra high stability quartz crystal oscillators. Electronics Letters. 2010;**46**(21): 1433-1434

[2] Volyanskiy K, Salzenstein P, Tavernier H, Pogurmirskiy M, Chembo YK, Larger L. Compact optoelectronic microwave oscillators using ultra-high q whispering gallery mode disk-resonators and phase modulation. Optics Express. 2010; **18**(21):22358-22363

[3] Curie J, Curie P. Développement par pression, de l'électricité polaire dans les cristaux hémièdres à faces inclinées. Comptes Rendus. 1880;**91**:294

[4] Salzenstein P. Recent progress in the performances of ultrastable Quartz resonators and oscillators. International Journal for Simulation and Multidisciplinary Design Optimization. 2016;7:A8(6)

Section 1

From Mathematics to Oscillators

Chapter 1

Mathematical Models of Oscillators with Memory

Roman Ivanovich Parovik

Abstract

The chapter proposes a mathematical model for a wide class of hereditary oscillators, which is a Cauchy problem in the local formulation. As an initial model equation, an integrodifferential equation of Voltaire type was introduced, which was reduced by means of a special choice of difference kernels to a differential equation with nonlocal derivatives of fractional-order variables. An explicit finite-difference scheme is proposed, and questions of its stability and convergence are investigated. A computer study of the proposed numerical algorithm on various test examples of the hereditary oscillators Airy, Duffing, and others was carried out. Oscillograms and phase trajectories are plotted and constructed.

Keywords: mathematical model, cauchy problem, heredity, derivative of fractional order, finite-difference scheme, stability, convergence, oscillograms, phase trajectory

1. Introduction

In the paper of the Italian mathematician Vito Volterra [1], the notion of heredity (memory), a property of a dynamical system characterized by nonlocality in time, is introduced, which consists in the dependence of its current state on a finite number of previous states. In another paper [2], Volterra investigated the hereditary oscillator—a vibration system with memory, which was written in the form of an integrodifferential equation with a difference kernel, a function of memory. Further, for such an oscillator, Volterra derived the law of total energy, in which an additional term appeared, responsible for the dissipation of energy in the vibrational system. This fact was confirmed in subsequent works.

In papers [3–21], fractal oscillators were considered, which represent the class of hereditary oscillators with a power-law function of memory. The peculiarity of such oscillators is that their mathematical description can be reduced to differential equations with nonlocal derivatives of fractional constant orders, which are investigated within the framework of the theory of fractional calculus [22–24].

In papers [6–9, 11–14, 18–21], models of fractal linear oscillators were investigated in the sense of the Gerasimov-Caputo derivative and in papers [4, 10, 16]—in the sense of the Riemann-Liouville derivative. Analytical solutions of model equations in terms of a special function of Mittag-Leffler-type and generalized Wright-type function, oscillograms, and phase trajectories are constructed. It is shown that in the regime of free oscillations, the presence of memory effects in the system leads to attenuation of oscillations as a result of energy dissipation, and with allowance for external periodic action, it is possible to stabilize the amplitude of the oscillations, with the phase trajectories reaching the limit cycle and also the resonance effect.

In papers of the author [25, 26], the fractal parametric resonance (the fractal Mathieu oscillator) was investigated, and the Strutt-Ince diagrams of parametric resonance existence areas were constructed. It is shown that these regions strongly depend on the orders of the value fractional derivatives entering into the initial equation.

In a monograph by the Slovak mathematician Ivo Petras [10], the fractal nonlinear oscillator models whose differential equations contained fractional derivatives in the sense of Riemann-Liouville were considered and analyzed using numerical methods and considered the stability of the rest point of oscillatory systems. However, the stability and convergence of numerical methods have not been considered.

A further continuation of the investigation of hereditary oscillators is associated with the introduction of the derivatives of fractional variable orders in the model equations. This is due to the fact that the orders of fractional derivatives are related to the properties of the medium in which this or that process takes place and changes with time under the influence of external influence. Therefore, papers [27–30] proposed that the models of fractal nonlinear oscillators (Duffing, Van der Pol, Van der Pol-Duffing, FitzHugh-Nagumo) were proposed and investigated using explicit finite-difference schemes, whose equations contain both the derivatives of the constants and variable fractional orders of the Gerasimov-Caputo and Riemann-Liouville types. With the help of computer experiments, the convergence of finite-difference schemes was shown, and estimates of the computational accuracy of the method were obtained; oscillograms and phase trajectories were constructed. However, the questions of stability and convergence were not formulated in the form of corresponding theorems.

In [31, 32], a new class of fractal oscillators was proposed and investigated; active fractal oscillators (AFOs)—nonlinear oscillators with external influences, which include the fractional Riemann-Liouville integral—were investigated. Such oscillators are constructed on the basis of the scheme of the radioelectronic autogenerator with a fractional feedback circuit. Authors use the method of equivalent linearization to investigate AFOs and come to the conclusion that the self-oscillator is isochronous.

From the analysis of the above publications on the study of hereditary oscillator, we can conclude that the main tool for their study is numerical methods, for example, finite-difference schemes. In most cases, the authors leave without considering the questions of stability and convergence of finite-difference schemes and, even if they touch, then without formulating the corresponding theorems and proofs. Therefore, the goal of the present paper is to construct a finite-difference scheme for a wide class of hereditary (fractal) linear and nonlinear oscillators, prove its stability and convergence, formulate results in the form of corresponding theorems, and study finite-difference schemes on specific test examples.

2. Formulation of the problem

Consider the following model integrodifferential equation for the function $x(t) \in C^3(0,T)$, where $T > 0$:

$$\int_0^t K_1(t-\eta)\ddot{x}(\eta)d\eta + \lambda\int_0^t K_2(t-\eta)\dot{x}(\eta)d\eta = f(x(t),t), \tag{1}$$

where $\ddot{x}(t) = d^2x/dt^2, \dot{x}(t) = dx/dt$, $\lambda > 0$—given constant.

Eq. (1) describes a wide class of hereditary, depending on the form of the right-hand side (function $f(x(t),t)$) of linear or nonlinear oscillators.

Definition 1. Functions $K_1(t-\eta)$ and $K_2(t-\eta)$—difference kernels in Eq. (1)—will be called memory functions, since they define the notion of heredity (memory), which was introduced in the work of the Italian mathematician Vito Volterra [2].

Definition 2. A nonlinear function $f(x(t),t)$ on the right-hand side of Eq. (1) satisfies a Lipschitz condition with respect to a variable $x(t)$:

$$|f(x_1(t),t) - f(x_2(t),t)| \leq L|x_1(t) - x_2(t)|, \tag{2}$$

L—Lipschitz constant.

Eq. (1) describes a broad class of hereditary nonlinear oscillators, depending on the form of the function $f(x(t),t)$ on its right-hand side, and the parameter λ has the meaning of the coefficient of friction.

Note that in **Definition 1**, the memory functions $K_1(t-\eta)$ and $K_2(t-\eta)$ can be chosen arbitrarily, depending on the conditions of the particular problem. We will choose these functions power law, since power laws are often found in various fields of knowledge [33]. We choose the memory functions $K_1(t-\eta)$ and $K_2(t-\eta)$ in the form

$$K_1(t-\eta) = \frac{(t-\eta)^{1-\beta(t)}}{\Gamma(2-\beta(t))}, K_2(t-\eta) = \frac{(t-\eta)^{-\gamma(t)}}{\Gamma(2-\gamma(t))}, 1 < \beta(t) < 2, 0 < \gamma(t) < 1, \tag{3}$$

where $\gamma(t), \beta(t) \in C[0,T]$, $\Gamma(t)$—Euler gamma function.

We give the following definitions.

Definition 3. Derivatives of fractional variables of orders $\beta(t)$ and $\gamma(t)$ Gerasimov-Caputo type: we call the following operators of fractional differentiation:

$$\partial_{0t}^{\beta(t)}x(\eta) = \frac{1}{\Gamma(2-\beta(t))}\int_0^t \frac{\ddot{x}(\eta)d\eta}{(t-\eta)^{\beta(t)-1}}, \partial_{0t}^{\gamma(t)}x(\eta) = \frac{1}{\Gamma(1-\gamma(t))}\int_0^t \frac{\dot{x}(\eta)d\eta}{(t-\eta)^{\gamma(t)}}. \tag{4}$$

We note that in the case when the functions $\beta(t)$ and $\gamma(t)$ in the relations (4) and (5) are constants, we arrive at the definitions of the fractional derivative in the sense of Gerasimov-Caputo [34, 35], and in the case when these constants $\beta = 2$ and $\gamma = 1$, the operators of fractional differentiation (4) become classical derivatives of the second and first orders.

Taking into account **Definition 3**, the model Eq. (1) can be rewritten in a more compact form:

$$\partial_{0t}^{\beta(t)}x(\eta) + \lambda\partial_{0t}^{\gamma(t)}x(\eta) = f(x(t),t). \tag{5}$$

For Eq. (5), the initial conditions in the local formulation are valid:

$$x(0) = \alpha_0, \dot{x}(0) = \alpha_1, \tag{6}$$

where α_0 and α_1—given constants. As a result, we arrive at the Cauchy problems (5) and (6), which we will investigate.

3. Explicit finite-difference scheme

We construct an explicit finite-difference scheme for the Cauchy problems (5) and (6). We divide the time interval $[0, T]$ into N equal parts with a step $\tau = T/N$. We introduce the grid function $x(t_k) = x_k$, where $t_k = k\tau, k = 1, ..., N-1$.

The derivatives of the variables of fractional orders in Eq. (5) are approximated according to the relations in [36, 37]:

$$\partial_{0t}^{\beta(t)} x(\eta) \approx A_k \sum_{j=0}^{k-1} a_{j,k} \left(x_{k-j+1} - 2x_{k-j} + x_{k-j-1}\right), \tag{7}$$

$$\partial_{0t}^{\gamma(t)} x(\eta) \approx B_k \sum_{j=0}^{k-1} b_{j,k} \left(x_{k-j+1} - x_{k-j-1}\right),$$

then the formulas will refer to the formula (7)

$$a_{j,k} = (j+1)^{2-\beta_k} - j^{2-\beta_k}, b_{j,k} = (j+1)^{1-\gamma_k} - j^{1-\gamma_k},$$

$$A_k = \frac{\tau^{-\beta_k}}{\Gamma(3-\beta_k)}, B_k = \frac{\lambda \tau^{-\gamma_k}}{2\Gamma(2-\gamma_k)},$$

Here, to shorten the record, $\beta(t_k) = \beta_k, \gamma(t_k) = \gamma_k$.

Taking into account relation (7), the Cauchy problems (5) and (6) in the difference formulation will have the form

$$A_k \sum_{j=0}^{k-1} a_{j,k} \left(x_{k-j+1} - 2x_{k-j} + x_{k-j-1}\right) + B_k \sum_{j=0}^{k-1} b_{j,k} \left(x_{k-j+1} - x_{k-j-1}\right) = f_k, \quad x_0 = \alpha_0, x_1 = \alpha_1 + \tau\alpha_0, \tag{8}$$

Here, to shorten the record, $f_k = f(x_k, t_k)$. We write the problem (8) explicitly:

$$x_{k+1} = \frac{1}{A_k + B_k} \left(2A_k x_k - (A_k - B_k) x_{k-1} - A_k \sum_{j=1}^{k-1} a_{j,k} \left(x_{k-j+1} - 2x_{k-j} + x_{k-j-1}\right) \right) - \frac{B_k}{A_k + B_k} \sum_{j=1}^{k-1} b_{j,k} \left(x_{k-j+1} - x_{k-j-1}\right) + f_k. \tag{9}$$

We note that the weight coefficients $a_{j,k}$ and $b_{j,k}$ have properties, which we formulate in the form of the following lemmas.

Lemma 1. *For any k weights, coefficients $a_{j,k}$ and $b_{j,k}$, as well as coefficients A_k and B_k, have the following properties:*

1. $\sum_{j=0}^{k-1} a_{j,k} = k^{2-\beta_k}, \sum_{j=0}^{k-1} b_{j,k} = k^{1-\gamma_k}$,

2. $1 = a_{0,k} > a_{1,k} > ... > 0, 1 = b_{0,k} > b_{1,k} > ... > 0$,

3. $A_k \geq 0, B_k \geq 0$.

Proof. The first property follows the definition of weight coefficients:

$$\sum_{j=0}^{k-1} a_{j,k} = \sum_{j=0}^{k-1} \left[(j+1)^{2-\beta_k} - j^{2-\beta_k} \right] = 1 - 0 + 2^{2-\beta_k} - 1 + 3^{2-\beta_k} - 2^{2-\beta_k} + ... + (k-1)^{2-\beta_k} + k^{2-\beta_k} - (k-1)^{2-\beta_k} = k^{2-\beta_k}.$$

$$\sum_{j=0}^{k-1} b_{j,k} = \sum_{j=0}^{k-1}\left[(j+1)^{1-\gamma_k} - j^{1-\gamma_k}\right] = 1 - 0 + 2^{1-\gamma_k} - 1 + 3^{1-\gamma_k} - 2^{1-\gamma_k} + \ldots.$$
$$+(k-1)^{1-\gamma_k} + k^{1-\gamma_k} - (k-1)^{1-\gamma_k} = k^{1-\gamma_k}.$$

The second property is proven in the following way. We introduce two functions:

$$\varphi(x) = (x+1)^{2-\beta_k} - x^{2-\beta_k} \text{ and } \eta(x) = (x+1)^{1-\gamma_k} - x^{1-\gamma_k},$$

where $x > 0$. These functions are decreasing. Really derived from these functions

$$\varphi'(x) = (2-\beta_k)\left[(x+1)^{1-\beta_k} - x^{1-\beta_k}\right] < 0, \eta'(x) = (1-\gamma_k)\left[(x+1)^{1-\gamma_k} - x^{1-\gamma_k}\right] < 0.$$

Therefore, the second property holds. The third property follows also the properties of the gamma function. The lemma is proven.

Let $\overline{\partial}_{0t}^{\beta(t)} x(\eta)$ and $\overline{\partial}_{0t}^{\gamma(t)} x(\eta)$—approximations of differential operators of Gerasimov-Caputo types $\partial_{0t}^{\beta(t)} x(\eta)$ and $\partial_{0t}^{\gamma(t)} x(\eta)$. Then, we have the following lemma.

Lemma 2. Approximations $\overline{\partial}_{0t}^{\beta(t)} x(\eta)$ and $\overline{\partial}_{0t}^{\gamma(t)} x(\eta)$ operators of the Gerasimov-Caputo types $\partial_{0t}^{\beta(t)} x(\eta)$ and $\partial_{0t}^{\gamma(t)} x(\eta)$ satisfy the following estimates:

$$\left|\partial_{0t}^{\beta(t)} x(\eta) - \overline{\partial}_{0t}^{\beta(t)} x(\eta)\right| \le C_1\tau, \left|\partial_{0t}^{\gamma(t)} x(\eta) - \overline{\partial}_{0t}^{\gamma(t)} x(\eta)\right| \le C_2\tau, \tag{10}$$

where C_1 and C_2—constants that are independent of the parameter τ.

Proof. Using the first property of **Lemma 1** and **Definition 3**, we obtain

$$\overline{\partial}_{0t}^{\beta(t)} x(\eta) = \frac{\tau^{2-\beta_k}}{\Gamma(3-\beta_k)}\sum_{j=0}^{k-1} a_{j,k}\left[\ddot{x}(t-j\tau) + O(\tau^2)\right] = \frac{\tau^{2-\beta_k}}{\Gamma(3-\beta_k)}\sum_{j=0}^{k-1} a_{j,k}\ddot{x}(t-j\tau)$$
$$+\frac{\tau^{2-\beta_k}k^{2-\beta_k}}{\Gamma(3-\beta_k)}O(\tau^2) = \frac{\tau^{2-\beta_k}}{\Gamma(3-\beta_k)}\sum_{j=0}^{k-1} a_{j,k}\ddot{x}(t-j\tau) + \frac{t^{2-\beta_k}}{\Gamma(3-\beta_k)}O(\tau^2)$$
$$= \frac{\tau^{2-\beta_k}}{\Gamma(3-\beta_k)}$$
$$\sum_{j=0}^{k-1} a_{j,k}\ddot{x}(t-j\tau) + O(\tau^2).$$

$$\partial_{0t}^{\beta} x(\eta) = \frac{1}{\Gamma(2-\beta_k)}\sum_{j=0}^{k-1}\int_{j\tau}^{(j+1)\tau} \xi^{1-\beta_k}\ddot{x}(t-\xi)d\xi = \frac{1}{\Gamma(2-\beta_k)}\sum_{j=0}^{k-1} a_{j,k}\ddot{x}\left(t-\eta_j\right),$$
$$\eta_j \in [j\tau, (j+1)\tau].\left|\partial_{0t}^{\beta(t)} x(\eta) - \overline{\partial}_{0t}^{\beta(t)} x(\eta)\right|$$
$$= \left|\frac{\tau^{2-\beta(t)}}{\Gamma(3-\beta(t))}\sum_{j=0}^{k-1} a_{j,k}\left[\ddot{x}(t-j\tau) - \ddot{x}\left(t-\eta_j\right)\right] + O(\tau^2)\right|$$
$$= \left|\frac{\tau^{2-\beta_k}}{\Gamma(3-\beta_k)}\sum_{j=0}^{k-1} a_{j,k}\cdot O(\tau) + O(\tau^2)\right| = \left|\frac{\tau^{2-\beta_k}k^{2-\beta_k}}{\Gamma(3-\beta_k)}O(\tau) + O(\tau^2)\right|$$
$$= O(\tau) + O(\tau^2) = O(\tau).$$

Similarly, we can obtain the second estimate from Eq. (10).

$$\overline{\partial}_{0t}^{\gamma(t)}x(\eta) = \frac{\tau^{1-\gamma_k}}{\Gamma(2-\gamma_k)}\sum_{j=0}^{k-1} b_{j,k}[\dot{x}(t-j\tau)+O(\tau)] = \frac{\tau^{1-\gamma_k}}{\Gamma(2-\gamma_k)}\sum_{j=0}^{k-1} b_{j,k}\,\dot{x}(t-j\tau) + \frac{\tau^{1-\gamma_k}k^{1-\gamma_k}}{\Gamma(2-\gamma_k)}$$

$$O(\tau) = \frac{\tau^{1-\gamma_k}}{\Gamma(2-\gamma_k)}\sum_{j=0}^{k-1} b_{j,k}\,\dot{x}(t-j\tau) + \frac{t^{1-\gamma_k}}{\Gamma(2-\gamma_k)}O(\tau^2) = \frac{\tau^{1-\gamma_k}}{\Gamma(2-\gamma_k)}\sum_{j=0}^{k-1} b_{j,k}\,\dot{x}(t-j\tau) + O(\tau).$$

$$\partial_{0t}^{\gamma(t)}x(\eta) = \frac{1}{\Gamma(1-\gamma_k)}\sum_{j=0}^{k-1}\int_{j\tau}^{(j+1)\tau}\xi^{-\gamma_k}\dot{x}(t-\xi)d\xi = \frac{1}{\Gamma(1-\gamma_k)}\sum_{j=0}^{k-1} b_{j,k}\dot{x}\left(t-\eta_j\right),$$

$$\eta_j \in [j\tau,(j+1)\tau].$$

$$\left|\partial_{0t}^{\gamma(t)}x(\eta) - \overline{\partial}_{0t}^{\gamma(t)}x(\eta)\right| = \left|\frac{\tau^{1-\gamma_k}}{\Gamma(2-\gamma_k)}\sum_{j=0}^{k-1} b_{j,k}\left[\dot{x}(t-j\tau) - \dot{x}\left(t-\eta_j\right)\right] + O(\tau)\right| =$$

$$= \left|\frac{\tau^{1-\gamma_k}}{\Gamma(2-\gamma_k)}\sum_{j=0}^{k-1} b_{j,k}\cdot O(\tau) + O(\tau)\right| = \left|\frac{\tau^{1-\gamma_k}k^{1-\gamma_k}}{\Gamma(2-\gamma_k)}O(\tau) + O(\tau)\right|$$

$$= O(\tau) + O(\tau) = O(\tau).$$

The lemma is proven.

Investigation. According to **Lemma 2**, it can be shown that the explicit finite-difference scheme (9) has an error $\varepsilon = O(\tau)$. This fact will be used in computer experiments in determining the computational accuracy of the numerical method.

Lemma 3. *The sums in the finite-difference scheme (9) have the following representations:*

$$\sum_{j=1}^{k-1} a_{j,k}\left(x_{k-j+1} - 2x_{k-j} + x_{k-j-1}\right) = a_{1,k}x_k + a_{k-1,k}x_0$$
$$+\left(a_{k-2,k} - 2a_{k-1,k}\right)x_1 + \left(a_{2,k} - 2a_{1,k}\right)x_{k-1} + \sum_{j=2}^{k-2}\left(a_{j+1,k} - 2a_{j,k} + a_{j-1,k}\right)x_{k-j},$$
$$\sum_{j=1}^{k-1} b_{j,k}\left(x_{k-j+1} - x_{k-j-1}\right) = b_{1,k}x_k - b_{k-1,k}x_0 + b_{2,k}x_{k-1} - b_{k-2,k}x_1$$
$$+\sum_{j=2}^{k-2}\left(b_{j+1,k} - b_{j-1,k}\right)x_{k-j}. \tag{11}$$

Proof. The representation (11) follows the properties of the sum. Indeed, by opening the sums in Eq. (9) and grouping the terms properly, we arrive at the representation (11).

Using **Lemma 3**, the finite-difference scheme (9) can be rewritten as

$$x_1 = \alpha_0 + \tau\alpha_1,$$

$$x_2 = \frac{2A_1}{A_1+B_1}x_1 - \frac{A_1-B_1}{A_1+B_1}x_0 + \frac{f_1}{A_1+B_1}, k=1,$$

$$x_{k+1} = \frac{1}{A_k+B_k}$$
$$\left[(A_k(2-a_{1,k}) - B_k b_{1,k})x_k - (A_k(a_{2,k} - 2a_{1,k} + a_{0,k}) + B_k(b_{2,k} - b_{0,k}))x_{k-1} + f_k\right] -$$
$$-\frac{(A_k(a_{k-2,k} - 2a_{k-1,k}) - B_k b_{k-2,k})x_1 - (A_k a_{k-1,k} - B_k b_{k-1,k})x_0}{A_k+B_k}$$
$$-\frac{1}{A_k+B_k}\sum_{j=2}^{k-2}\left[A_k\left(a_{j+1,k} - 2a_{j,k} + a_{j-1,k}\right) + B_k\left(b_{j+1,k} - b_{j-1,k}\right)\right]x_{k-j}, k = 2, \ldots, n-1,$$

Or in matrix form

$$X_{k+1} = MX_k + F_k, \tag{12}$$

$$X_{k+1} = (x_1, x_2, \dots, x_{N-1})^T, X_k = (x_0, x_1, \dots, x_{N-2})^T, F_k = (f_0, f_1, \dots, f_{N-2})^T$$

where the matrix $M = (m_{ij})$, $i = 1, \dots, N-1, j = 1, \dots, N-1$:

$$m_{ij} = \begin{cases} 0, & j \geq i+1, \\ \dfrac{A_{i-1}(2 - a_{i-2,i-1}) - B_{i-1}b_{i-2,i-1}}{A_{i-1} + B_{i-1}}, & j = i = 3, \dots, N-1, \\ \dfrac{-A_{i-1}(a_{i-j+1,i-1} - 2a_{i-j,i-1} + a_{i-j-1,i-1}) - B_{i-1}(b_{i-j+1,i-1} - b_{i-j-1,i-1})}{A_{i-1} + B_{i-1}}, & j \leq i-1, \end{cases} \tag{13}$$

$$m_{1,1} = 1, m_{2,2} = \frac{2A_1}{A_1 + B_1}, m_{i,1} = \frac{B_{i-1}b_{i-2,i-1} - A_{i-1}a_{i-2,i-1}}{A_{i-1} + B_{i-1}}, i = 2, \dots, N-1,$$

$$m_{i,2} = \frac{A_{i-1}(2a_{i-2,i-1} - a_{i-3,i-1}) + B_{i-1}b_{i-3,i-1}}{A_{i-1} + B_{i-1}}, i = 3, \dots, N-1.$$

Theorem 1. *An explicit finite-difference scheme (9) converges with the first order* $|\overline{x}_k - x_k| = O(\tau)$ *if the following condition is satisfied:*

$$\tau \leq \tau_0 = \min\left(1, \left(\frac{2\Gamma(2-\gamma_{i-1})}{\lambda\Gamma(3-\beta_{i-1})}\right)^{\frac{1}{\beta_{i-1}-\gamma_{i-1}}}\right), i = 2, \dots, N-1. \tag{14}$$

Proof. Let $\overline{X}_k = (\overline{x}_0, \dots, \overline{x}_{N-2})^T$ be the exact solution of system (8) and the error vector $e_{k+1} = \overline{X}_{k+1} - X_{k+1}, e_0 = 0$.. Then, system (8), with allowance for **Lemma 2**, can be written as follows:

$$e_{k+1} = Me_k + F_{e,k} + O(\tau), \tag{15}$$

where

$$F_{e,k} = \frac{1}{A_k + B_k}(|f(x_1, t_k) - f(\overline{x}_1, t_k)|, \dots, |f(x_k, t_k) - f(\overline{x}_k, t_k)|)^T$$
$$\leq \frac{1}{A_k + B_k}(L_1 e_1, \dots, L_k e_k) = \Delta F_k e_k, \Delta F_k = \frac{1}{A_k + B_k} diag(L_1, \dots, L_k)^T.$$

We note that for any k, the inequality holds $|L_k| \leq L$. Consider the norm for the matrix M:$\|M\|_\infty = \max_i\left(\sum_{j=1}^{k-1} m_{ij}\right)$, we obtain

$$\|M\|_\infty = \max_{1 \leq i \leq N-1} \left(\begin{array}{l} 1 + \left|\dfrac{B_1 b_{0,1} - A_1 a_{0,1}}{A_1 + B_1}\right| + \left|\dfrac{2A_1}{A_1 + B_1}\right| + \left|\dfrac{B_2 b_{1,2} - A_2 a_{1,2}}{A_2 + B_2}\right| \\ + \left|\dfrac{A_2(2a_{1,2} - a_{0,2}) + B_2 b_{0,2}}{A_2 + B_2}\right| + \left|\dfrac{A_2(2 - a_{1,2}) - B_2 b_{1,2}}{A_2 + B_2}\right| \\ + \left|\dfrac{A_3(2a_{1,3} - a_{2,3} - a_{0,3}) + B_3(b_{0,3} - b_{2,3})}{A_3 + B_3}\right| + \dots + \left|\dfrac{B_{i-1}b_{i-2,i-1} - A_{i-1}a_{i-2,i-1}}{A_{i-1} + B_{i-1}}\right| \\ + \left|\dfrac{A_{i-1}(2a_{i-2,i-1} - a_{i-3,i-1}) + B_{i-1}b_{i-3,i-1}}{A_{i-1} + B_{i-1}}\right| \\ + \left|\dfrac{A_{i-1}(2a_{i-3,i-1} - a_{i-2,i-1} - a_{i-4,i-1}) + B_{i-1}(b_{i-4,i-1} - b_{i-2,i-1})}{A_{i-1} + B_{i-1}}\right| \\ + \left|\dfrac{A_{i-1}(2 - a_{i-2,i-1}) - B_{i-1}b_{i-2,i-1}}{A_{i-1} + B_{i-1}}\right| + \dots \end{array} \right) \tag{16}$$

According to **Lemma 1**, we note that the inequality holds $\frac{2A_1}{A_1+B_1} \geq 0$. Suppose that the condition is satisfied $A_{i-1} \geq B_{i-1}$, then the second diagonal element satisfies the inequality $1 \leq \frac{2A_1}{A_1+B_1} \leq 2$ in the matrix M, and the remaining diagonal elements are equal to the inequality $0 \leq \frac{A_{i-1}(2-a_{i-2,i-1})-B_{i-1}b_{i-2,i-1}}{A_{i-1}+B_{i-1}} \leq 1$; these elements with $i \to N-1$, in view of properties 2 and 3, tend to be zero.

The remaining elements of the matrix (16) also possess these properties. We also note that the matrix M is a matrix with a diagonal predominance for small values λ.

Therefore, the sum of the elements of the second row in the matrix M satisfies the condition $1 \leq \frac{2A_1}{A_1+B_1} + \frac{A_1 a_{0,1} - B_1 b_{0,1}}{A_1+B_1} \leq 3$. Further, by virtue of properties 2 and 3 of **Lemma 1**, it is obvious that the sum of the remaining terms also satisfies these conditions. Therefore, the following estimate is valid for the norm: $1 \leq \|M\|_\infty \leq 3$.

Note that for the values of the parameter $\lambda \gg 1$ the norm $\|M\|_\infty \to 1$, however, the condition number $\mu(M) \gg 1$ is violated and the diagonal transformation is violated; therefore, it is necessary to decrease the step τ.

Further, for any constant $C > 0$ independent of τ, and the error rate, the following estimate holds:

$$\|e_{k+1}\|_\infty \leq \|\Delta F_k + M\|_\infty \|e_k\|_\infty + C\tau \leq \left(3 + \frac{L}{A_k + B_k}\right)\|e_k\|_\infty + C\tau. \tag{17}$$

We introduce the notation in Eq. (17): $s_k = \left(3 + \frac{L}{A_k+B_k}\right)$, $s = C\tau$. Then, we obtain the following estimate:

$$\begin{aligned}
\|e_{k+1}\|_\infty &\leq s_k\|e_k\|_\infty + s \leq s_k\left(s_{k-1}\|e_{k-1}\|_\infty + s\right) + s = s_k s_{k-1}\|e_{k-1}\|_\infty + s(s_k+1) \\
&\leq s_k s_{k-1}\left(s_{k-2}\|e_{k-2}\|_\infty + s\right) + s(s_k+1) = s_k s_{k-1} s_{k-2}\|e_{k-2}\|_\infty + s(s_k s_{k-1} + s_k + 1) \\
&\leq s_k s_{k-1} s_{k-2}\left(s_{k-3}\|e_{k-3}\|_\infty + s\right) + s(s_k s_{k-1} + s_k + 1) = s_k s_{k-1} s_{k-2} s_{k-3}\|e_{k-3}\|_\infty \\
&\leq s_k s_{k-1} s_{k-2}\left(s_{k-3}\|e_{k-3}\|_\infty + s\right) + s(s_k s_{k-1} + s_k + 1) = s_k s_{k-1} s_{k-2} s_{k-3}\|e_{k-3}\|_\infty \\
&\quad + s(s_k s_{k-1} s_{k-2} + s_k s_{k-1} + s_k + 1) \leq \ldots \leq s_k s_{k-1} \cdot \ldots \cdot s_{k-r}\|e_{k-r}\| \\
&\quad + s(s_k s_{k-1} \cdot \ldots \cdot s_{k-r+1} + \ldots + s_k + 1).
\end{aligned} \tag{18}$$

Substituting into Eq. (18) $r = k-1$, we obtain

$$\|e_{k+1}\|_\infty \leq s_k s_{k-1} \cdot \ldots \cdot s_1\|e_1\| + s(s_k s_{k-1} \cdot \ldots \cdot s_2 + \ldots + s_k + 1) \leq C_0\|e_0\| + O(\tau).$$

From the second initial condition (6) it follows: $\|e_1\| \leq \|e_0\|$ and $C_0 = \prod_{p=1}^{k} s_p$.

Now according to our assumption $A_{i-1} \geq B_{i-1}$, which leads us to the relation

$$\tau \leq \left(\frac{2\Gamma(2-\gamma_{i-1})}{\lambda\Gamma(3-\beta_{i-1})}\right)^{\frac{1}{\beta_{i-1}-\gamma_{i-1}}}, i = 2, \ldots, N-1. \tag{19}$$

Condition (19) begins to work at such values λ, when many of conditionalities arise $\mu(M) \gg 1$, and for sufficiently small values λ, it suffices that the step satisfies the inequality $\tau \leq 1$. Therefore, we arrive at the relation (14). The theorem is proven.

We note that in [38] the authors used the classical Lax theorem, which holds for local finite-difference schemes, to prove the convergence of the scheme. For nonlocal finite-difference schemes, the convergence must be proven independently.

We consider the stability of an explicit finite-difference scheme (4). Suppose that X_k and two Y_k are different solutions of the matrix Eq. (12) with initial conditions X_0 and Y_0.

Theorem 2. *An explicit finite-difference scheme (9) is conditionally stable if condition (14) is satisfied and the estimate holds* $|Y_k - X_k| \le C|Y_0 - X_0|$ *for any k, where* $C > 0$ *does not depend on the step* τ.

Proof. We introduce the notation: $e_{k+1} = Y_{k+1} - X_{k+1}$. Then, Eq. (12) can be written in the form $e_{k+1} = Me_k + F_{e,k}$. Here, as it was said in.

$F_{e,k} = \frac{1}{A_k+B_k}(|f(x_1,t_k) - f(\overline{x}_1,t_k)|, ..., |f(x_k,t_k) - f(\overline{x}_k,t_k)|)^T \le \frac{1}{A_k+B_k}$ $(L_1e_1, ..., L_ke_k) = \Delta F_k e_k$

According to **Theorem 1**, we have the following estimate:

$$\|M + \Delta F_k\| \le \left(3 + \frac{L}{A_k + B_k}\right) = s_k.$$

Therefore, the following estimate holds

$$\|e_{k+1}\|_\infty \le \|M + \Delta F_k\| \|e_k\|_\infty \le \left(3 + \frac{L}{A_k + B_k}\right) \|e_k\|_\infty$$

$$= s_k\|e_k\|_\infty \le s_k s_{k-1}\|e_{k-1}\|_\infty \le s_k s_{k-1} s_{k-2}\|e_{k-2}\|_\infty \le ... \le s_k s_{k-1} \cdot ... \cdot s_{k-r}\|e_{k-r}\|.$$

With $r = k - 1$, we obtain $\|e_{k+1}\|_\infty \le C_0\|e_1\| \le C_0\|e_0\|$ and $C_0 = \prod_{p=1}^{k} s_p$.

The last inequality follows the second condition of problem (6). Therefore, if X_0 there is a perturbation, then it does not lead to a large increase in the error of the numerical solution. However, for large values λ, many of conditionalities $\mu(M) \gg 1$ arise, and therefore it is necessary to decrease the step τ; according to Eq. (19), for small values λ, the estimate is valid $\tau \le 1$. Then, the system is stable if condition (14) is satisfied. The theorem is proven.

4. Results of modeling

Consider the work of the explicit finite-difference scheme (9) on specific examples. We show that the scheme (9) has the first order of accuracy. Since in the general case, the exact solution of the Cauchy problems (5) and (6) cannot be written in analytical form, we will use the double conversion method. For this, we introduce two parameters: $\xi = \max_i |x_i - x_{2i}|$—absolute error between the numerical solution x_i in step τ and the numerical solution x_{2i} in step $\tau/2$. Then, the order of computational accuracy p can be estimated by the formula

$$p = \log_2(\xi)/\log_2(\tau/2).$$

We note that in the case when the fractional parameters in the scheme (9) do not change and have the following values of $\beta_k = 2$ and $\gamma_k = 1$, we arrive at the classical local explicit finite-difference scheme with the second order of accuracy.

The numerical algorithm (9) was implemented in Maple software.

Example 1. Suppose that the right-hand side in Eq. (1) has the form

$$f(x(t), t) = \delta \sin(\varphi t) + tx(t).$$

Then, Eq. (5) describes a linear hereditary Airy oscillator, which was considered in the author's papers [21, 39] and has the following form.

$$\partial_{0t}^{\beta(t)}x(\eta) + \lambda\partial_{0t}^{\gamma(t)}x(\eta) - tx(t) = \delta\cos(\varphi t).$$

We choose the initial condition (6) for simplicity by homogeneous:

$$x(0) = \dot{x}(0) = 0.$$

We note that the Airy oscillator is used in optics in the simulation of Airy laser beams. In this case, the explicit finite-difference scheme (9) has a more specific form:

$$x_0 = x_1 = 0,$$

$$x_{k+1} = \frac{1}{A_k + B_k}((2A_k - k\tau)x_k - (A_k - B_k)x_{k-1}) \qquad (20)$$

$$-\frac{A_k}{A_k + B_k}\sum_{j=1}^{k-1} a_{j,k}\left(x_{k-j+1} - 2x_{k-j} + x_{k-j-1}\right)$$

$$-\frac{B_k}{A_k + B_k}\sum_{j=1}^{k-1} b_{j,k}\left(x_{k-j+1} - x_{k-j-1}\right) + \delta\sin(\varphi k\tau).$$

For the explicit finite-difference scheme (20), we choose the following values of the control parameters: $T = 1.$ $\lambda = 1$, $\delta = 5$, $\varphi = 10$, $\omega = 10$ and $\beta(t) = 1.8 - 0.03\sin(\omega t)$, $\gamma(t) = 0.8 - 0.05\cos(\omega t)$. And during the simulation, we will change the number of nodes N in the calculation grid.

Note that the values of the selected parameters for **Example 1** satisfy the conditions of **Theorems 1** and **2**, which is indirectly confirmed by the results of modeling for different values N of the nodes of the computational grid (**Table 1**).

From **Table 1** we can notice that when the number of calculated nodes in the grid doubles in nodes N, the maximum error in absolute value decreases twice, and the order of computational accuracy p tends to unite.

This confirms that the explicit finite-difference scheme (9) and in particular the scheme (20) for **Example 1** have the first order of accuracy, and since condition (14) is satisfied, then convergence with the same order.

In **Figure 1** the oscillogram (**Figure 1a**) and the phase trajectory (**Figure 1b**) are shown for **Example 1** at the parameter value $T = 10$, $N = 1000$. It can be noted that with time the amplitude of the oscillations is established and as a result the phase trajectory reaches the limit cycle. Another situation arises in the case of free oscillations $\delta = 0$ (**Figure 2**).

The amplitude of the oscillations decays (**Figure 2a**), and the phase trajectory twists into a spiral (**Figure 2b**). The dissipation of energy in this case occurs as a result of the presence of friction with a coefficient λ and also the "memory" effect, which gives an additional term in the ratio for the total energy of the oscillatory system (**Figure 3**).

This fact is confirmed by the results of [2]. Consider the following example of a nonlinear hereditary oscillator.

Example 2. Let that in Eq. (1) the right-hand side has the form.

$$f(x(t), t) = \delta\sin(\varphi t) - ax(t) + bx^3(t),$$

N	ξ	p
640	0.0003331017	1.119146497
1280	0.0001745618	1.102636795
2560	0.0000906971	1.089811915

Table 1.
Results of numerical simulation.

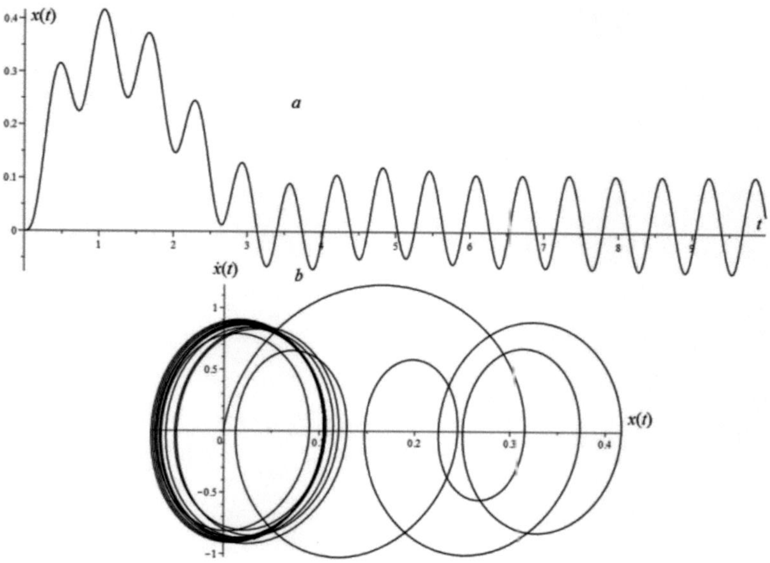

Figure 1.
The oscillogram (a) and the phase trajectory (b) for Example 1 with the parameter values $T = 10, N = 1000$.

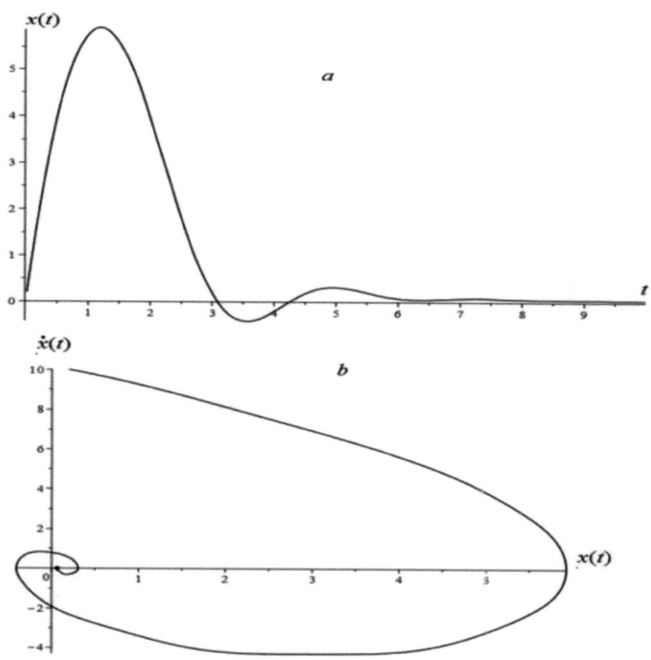

Figure 2.
Oscillogram (a) and phase trajectory (b) for Example 1 with initial conditions $x(0) = 0.1, \dot{x}(0) = 0.2$, *and* $\delta = 0$.

and we choose the initial conditions (6) to be homogeneous:

$$x(0) = \dot{x}(0) = 0.$$

In this case, Eq. (5) describes the Duffing fractional oscillator [18]:

$$\partial_{0t}^{\beta(t)} x(\eta) + \lambda \partial_{0t}^{\gamma(t)} x(\eta) + bx^3(t) - ax(t) = \delta \sin(\varphi t).$$

The explicit finite-difference scheme (9) for this case has the form

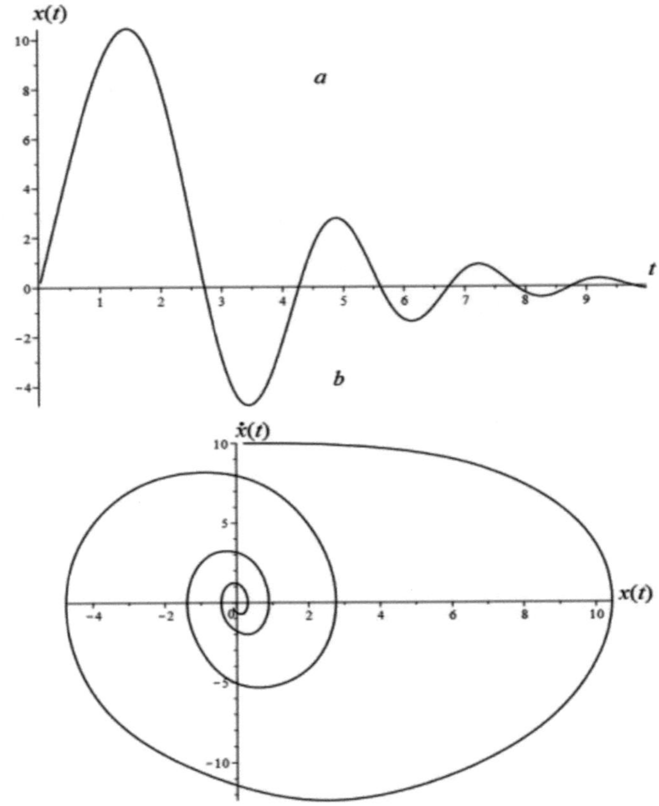

Figure 3.
The oscillogram (a) and the phase trajectory (b) for Example 1 with initial conditions $x(0) = 0.1, \dot{x}(0) = 0.2$, *and* $\delta = 0$, $\lambda = 0$.

N	ξ	p
640	0.0003619281	1.107545912
1280	0.0001896841	1.092050182
2560	0.0000991471	1.079382204

Table 2.
Results of numerical simulation.

$$x_0 = x_1 = 0,$$

$$x_{k+1} = \frac{1}{A_k + B_k}\left((2A_k + 1)x_k - x_k^3 - (A_k - B_k)x_{k-1}\right) \tag{21}$$
$$- \frac{A_k}{A_k + B_k}\sum_{j=1}^{k-1} a_{j,k}\left(x_{k-j+1} - 2x_{k-j} + x_{k-j-1}\right)$$
$$- \frac{B_k}{A_k + B_k}\sum_{j=1}^{k-1} b_{j,k}\left(x_{k-j+1} - x_{k-j-1}\right) + \delta\sin(\varphi k\tau).$$

For the explicit finite-difference scheme (21), we take the values of the control parameters as follows: $T = 1$, $\lambda = 0.3$, $\delta = 2$, and $\varphi = \omega = 1$.

Remark. Note that this choice of control parameter values is ensured by the condition (14) for **Theorems 1** and **2**. The results of numerical simulation for **Example 2** are given in **Table 2**.

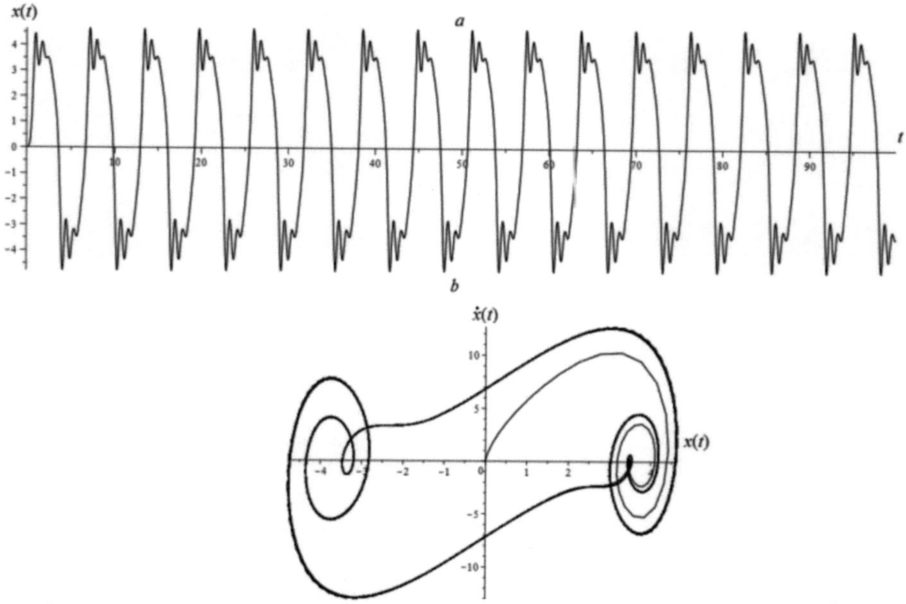

Figure 4.
The oscillogram (a) and the phase trajectory (b) for ***Example*** *2.*

Note from **Table 2** that for **Example 2**, with an increase in the number of design nodes N, the maximum error ξ in absolute value decreases and the order of computational accuracy p tends to unite. This indicates that the explicit finite-difference scheme (21) has the first order of accuracy.

Let's perform numerical simulation according to the scheme (21) with the values of the following parameters, $T = 100$, $N = 2000$, and $\delta = 50$, and leave the remaining parameters unchanged. Let us construct an oscillogram and a phase trajectory (**Figure 3**).

The oscillogram (**Figure 4a**) has a constant amplitude of a more complex shape at its minima and maxima, which is reflected in the phase trajectory (**Figure 4b**). The phase trajectory enters a complex two-loop limit cycle. The presence of such loops, apparently, is associated with the effects of memory in the oscillatory system [40].

Figure 5 shows the case of free oscillations for **Example 2**. It is seen that the presence of friction and memory effects in the oscillatory system intensify energy dissipation, which leads to damping of the oscillations (**Figure 5a**) and a phase trajectory—a twisting spiral (**Figure 5b**). Indeed, if there is no friction $\lambda = 0$ in the oscillatory system, we obtain an oscillogram and a phase trajectory as in **Figure 6**.

Example 3. Suppose that in Eq. (1) the right-hand side has the form

$$f(x(t),t) = bt + c\sum_{n=1}^{7} a_n \sin(nx(t)) - \omega^{\beta(t)}x(t), \tag{22}$$

where b is the spring travel speed; c is the surface adhesion energy; ω is the frequency of free oscillations; and $a_n = 2n\int_0^1 \frac{\cos(\pi n\tau)d\tau}{\cosh^2(\pi\tau)}$ is the coefficients of the expansion of the Fourier series.

Eq. (5) with the right-hand side of Eq. (22) describes the hereditary stick–slip effect [20]. The stick–slip effect is encountered in tribology problems, for example, when the movement of a load on a spring along a surface is investigated. Due to adhesion, the load adheres to the surface, and due to the tension of the spring, it

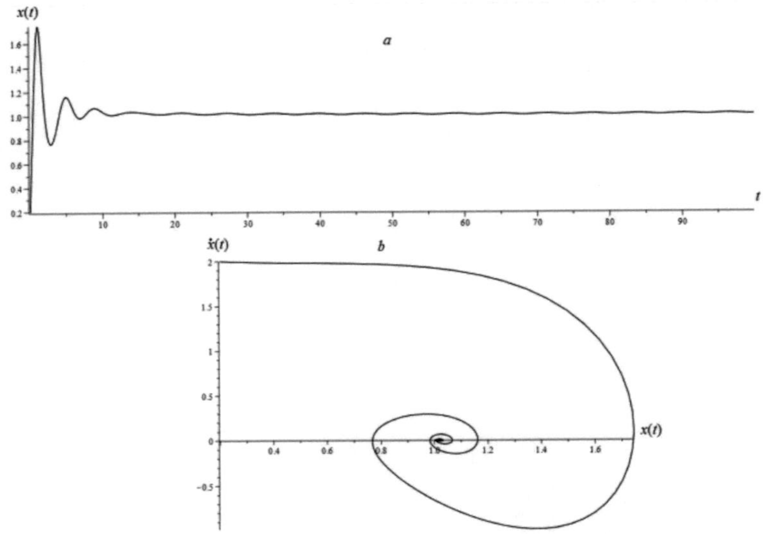

Figure 5.
Oscillogram (a) and phase trajectory (b) for **Example 3** *with initial conditions* $x(0) = 0.1, \dot{x}(0) = 0.2$*, and* $\delta = 0$.

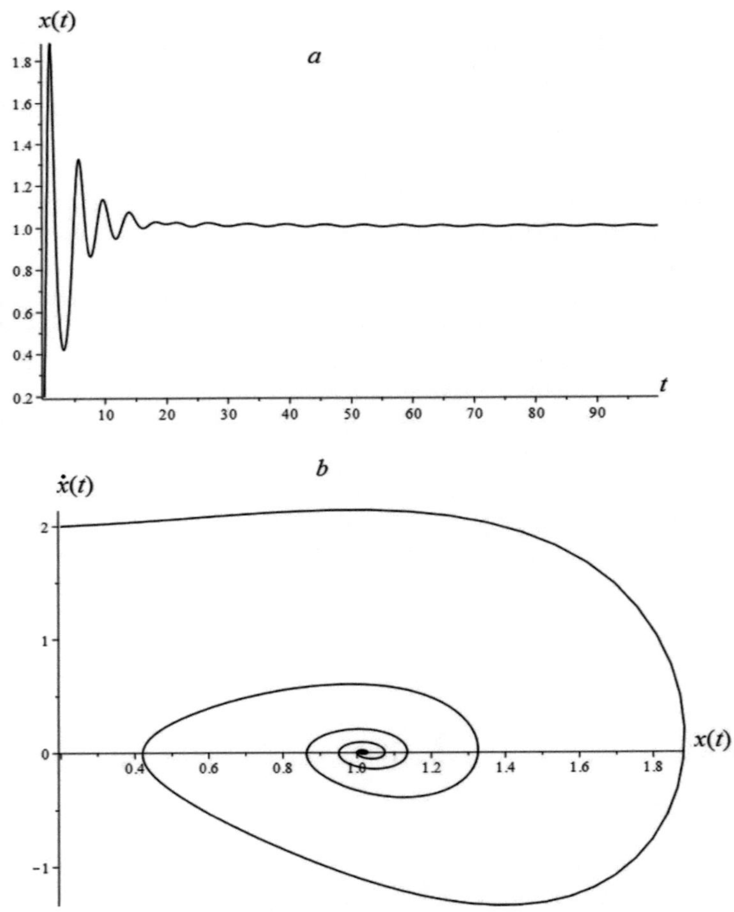

Figure 6.
Oscillogram (a) and phase trajectory (b) for **Example 2** *with initial conditions* $x(0) = 0.1, \dot{x}(0) = 0.2$*, and* $\delta = 0, \lambda = 0$.

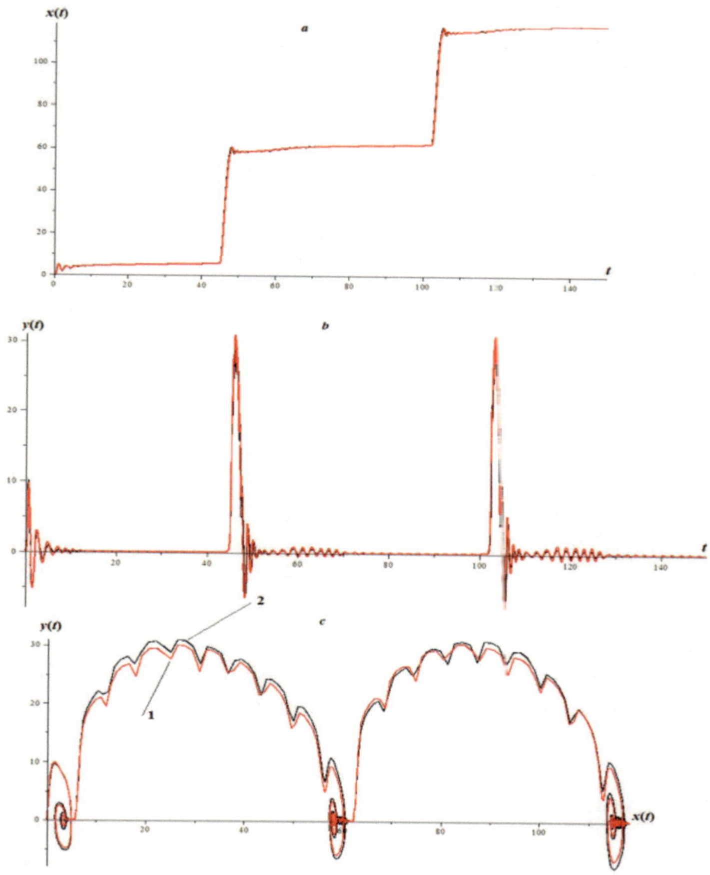

Figure 7.
Calculated curves obtained from formula (6) from [20] (curve 1) and formula (9) (curve 2): (a) oscillogram, (b) oscillator speed, and (c) phase trajectory.

breaks and slides along it, and its oscillations occur [41, 42]. The stick–slip effect can also be incorporated into the mechanical model of an earthquake in the subduction zone of lithospheres plates [43].

In [43] it was said that in order to obtain a reliable solution it suffices to take the first seven coefficients a_n in the expansion of the function (22). The values of these coefficients are taken from [43] $a_1 = 0.436$, $a_2 = 0.344$, $a_3 = 0.164$, $a_4 = 0.058$, $a_5 = 0.021$, $a_6 = 0.004$, and $a_7 = 0.003$. Values of control parameters are $\beta(t) = 1.8 - 0.03\sin(\pi t)$ $\gamma(t) = 0.6 - 0.04\cos(\pi t)$, $N = 3000$ $\delta = 50$, $\tau = 0.05$, $\lambda = 0.3$, $b = 1$, $\omega = 1$, and $x(0) = 0, \dot{x}(0) = 0.3$.

Figure 7 shows the calculated displacement curves, displacement velocities, and phase trajectory. **Figure 7a** shows the oscillogram for **Example 3**. It can be seen that during the separation, the load experiences oscillations and the rate of such oscillations in the potential well attenuates rather slowly (**Figure 7b**). This effect is the eradication of the process. The phase trajectory in **Figure 7c** shows that the potential wells are stable focuses.

5. Conclusion

A mathematical model characterizing a wide class of hereditary oscillators is proposed and studied. The model is a differential Cauchy problem with derivatives of fractional-order variables of the Gerasimov-Caputo types (5) and (6). Using the theory of finite-difference schemes, a nonlocal explicit finite-difference scheme (9) was constructed with the first order of accuracy. Questions of its stability and convergence, which are formulated in the form of corresponding theorems, were studied.

The main result of the paper can be formulated as follows: an explicit finite-difference scheme is conditionally stable and converges if criterion (14) is satisfied. With the help of computational examples, it was shown that the scheme (9) has the first order of accuracy. It is confirmed that in the case of free oscillations, the presence of friction and heredity increases dissipation of energy, which leads to attenuation of oscillations.

One of the continuations of the investigation of the Cauchy problems (5) and (6) is a generalization of it:

$$\partial_{0t}^{\beta(x(t),t)}x(\eta) + \lambda(x(t),t)\partial_{0t}^{\gamma(x(t),t)}x(\eta) = f(x(t),t), x(0) = \alpha_0, \dot{x}(0) = \alpha_1.$$

Another continuation of the research is related to the introduction of other memory functions $K_1(t-\tau)$, $K_2(t-\tau)$ into the model Eq. (1), which leads to different model equations with different derivatives of fractional orders, and also the Cauchy problems (5) and (6) can be written in terms of the local fractional derivative [44–46].

The question of the stability of the rest points of dynamical systems described by the Cauchy problems (5) and (6) is also interesting, by analogy with the papers [47, 48].

Acknowledgements

The work was carried out according to the state task within the framework of research work Vitus Bering Kamchatka State University on the topic "Application of fractional calculus in the theory of oscillatory processes" No.AAAA-A17-117031050058-9 and with the support of the grant of the President of the Russian Federation MK-1152.2018.1.

Author details

Roman Ivanovich Parovik[1,2*]

1 Institute of Cosmophysical Research and Radio Wave Propagation FEB RAS, v. Paratunka, Kamchatskiy kray, Russia

2 Vitus Bering Kamchatka State University, Petropavlovsk-Kamchatskiy, Russia

*Address all correspondence to: romanparovik@gmail.com

References

[1] Volterra V. Sur les équations intégro-différentielles et leurs applications. Acta Math. 1912;**35**(1):295-356

[2] Volterra V. Theory of Functionals and of Integral and Integro-differential Equations. New York: Dover Publication Inc; 2005. 226 p

[3] Mainardi F. Fractional relaxation-oscillation and fractional diffusion-wave. Chaos, Solitons & Fractals. 1996; 7(9):146-1477. DOI: 10.1016/0960-0779(95)00125-5

[4] Meilanov R, Yanpolov M. Features of the phase trajectory of a fractal oscillator. Technical Physics Letters. 2002;**28**(1):30-32. DOI: 10.1134/1.1448634

[5] Achar B, Hanneken J, Clarke T. Response characteristics of a fractional oscillator. Physica A: Statistical Mechanics and its Applications. 2002; **309**(3–4):275-288. DOI: 10.1016/S0378-4371(02)00609-X

[6] Nakhusheva V. Differential Equations of Mathematical Models of Non-Local Processes. Moscow: Nauka; 2006. p. 173

[7] Al-Rabtah A, Ertürk V, Momani S. Solutions of a fractional oscillator by using differential transform method. Computers & Mathematics with Applications. 2010;**59**(3):1356-1362. DOI: 10.1016/j.camwa.2009.06.036

[8] Rand R, Sah S, Suchrsky M. Fractional Mathieu equation. Communications in Nonlinear Science and Numerical Simulation. 2010;**15**: 3254-3262. DOI: 10.1016/j.cnsns.2009.12.009

[9] Afanas'ev V, Daniel M. Polish stabilization of the inertial effects of the fractal oscillator. Technical Physics Letters. 2010;**36**(7):1-6

[10] Petras I. Fractional-Order Nonlinear Systems. Modeling, Analysis and Simulation. Beijing and Springer-Verlag. Berlin Heidelberg: Springer; 2011. 218 p

[11] Zurigat M. Solving fractional oscillators using laplace homotopy analysis method. Annals of the University of Craiova, Mathematics and Computer Science Series. 2011;**38**(4): 1-11

[12] Gomez-Aguilar JF, Rosales-Garc'ıab JJ, Bernal-Alvaradoa JJ, Cordova-Fraga T, Guzman-Cabrera R. Fractional mechanical oscillators. Revista Mexicana de F'ısica. 2012;**58**:348-352

[13] Duan J-S. The periodic solution of fractional oscillation equation with periodic input. Adv. Math. Phys. 2013; **2013**:869484. DOI: 10.1155/2013/869484

[14] Parovik R. Fractal parametric oscillator as a model of a nonlinear oscillation system in natural mediums. International Journal of Communications, Network and System Sciences. 2013;**6**(3):134-138

[15] Xu Y, Agrawal O. Models and numerical solutions of generalized oscillator equations. Journal of Vibration and Acoustics. 2014;**136**:051005. DOI: 10.1115/1.4027241

[16] Syta A, Litak G, Lenci S, Scheffler M. Chaotic vibrations of the Duffing system with fractional damping. Chaos: An Interdisciplinary Journal of Nonlinear Science. 2014;**24**(1):013107. DOI: 10.1063/1.4861942

[17] Blaszczyk T. A numerical solution of a fractional oscillator equation in a non-resisting medium with natural boundary

conditions. Romanian Reports in Physics. 2015;**67**(2):350-358

[18] Parovik R. Mathematical modeling of nonlocal oscillatory Duffing system with fractal friction. Bulletin KRASEC. Physical and Mathematical Sciences. 2015;**10**(1):16-21. DOI: 10.18454/2313-0156-2015-10-1-16-21

[19] Parovik R. Mathematical modeling of the hereditary oscillator. Computer Research and Modeling. 2015;7(5): 1001-1021

[20] Parovik R. On a credit oscillatory system with the inclusion of stick-slip. E3S Web of Conferences. 2016;**11**: 00018. DOI: 10.1051/e3sconf/20161100018

[21] Parovik R. Mathematical modelling of hereditarity airy oscillator with friction. Bulletin of South Ural State University. Series Mathematical Modelling, Programming & Computer Software. 2017;**10**(1):138-148. DOI: 10.14529/mmp170109

[22] Nakhushev A. Fractional Calculus and its Applications. Moscow: Fizmatlit; 2003. 272 p

[23] Kilbas A, Srivastava H, Trujillo J. Theory and Applications of Fractional Differential Equations. Amsterdam: Elsevier; 2006. 523 p

[24] Uchaikin V. Fractional Derivatives for Physicists and Engineers. Vol. I. Background and Theory. Beijing/Berlin: Higher Education Press/Springer-Verlag; 2013. 373 p

[25] Parovik R. Phase trajectories of the fractal parametric oscillator. Journal of Engineering Research and Applications. 2013;**3**(5):1520-1523. DOI: 10.4236/ijcns.2013.63016

[26] Parovik R. Charts Strutt-Ince for generalized Mathieu equation. Bulletin KRASEC. Physical and Mathematical Sciences. 2012;**1**(4):30-34. DOI: 10.18454/2079-6641-2012-4-1-29-30

[27] Drobysheva I. Mathematical modeling of nonlinear hereditary oscillators on the example of Duffing oscillator with fractional derivatives in the sense of Riemann-Liouville. Bulletin KRASEC. Physical and Mathematical Sciences. 2016;**13**(2): 39-45. DOI: 10.18454/2313-0156-2016-13-2-39-45

[28] Kim V. Duffing oscillator with external harmonic action and variable fractional Riemann-Liouville derivative characterizing viscous friction. Bulletin KRASEC. Physical and Mathematical Sciences. 2016;**13**(2):46-49. DOI: 10.18454/2313-0156-2016-13-2-46-49

[29] Lipko O. Mathematical model of propagation of nerve impulses with regard hereditarity. Bulletin KRASEC. Physical and Mathematical Sciences. 2017;**17**(1):33-43. DOI: 10.18454/2313-0156-2017-16-1-52-60

[30] Novikova E. Van der Pol-Duffing oscillator with the effect of hereditary. Bulletin KRASEC. Physical and Mathematical Sciences. 2017;**17**(2): 65-75. DOI: 10.18454/2079-6641-2017-18-2-65-75

[31] Zaitsev V, Ar K, Yarovoy G. Self-oscillations dynamics of active fractional oscillator. Theoretical Physics. 2013;**14**: 11-18

[32] Zaitsev V, Karlov A, Yarovoy G. Self-oscillations dynamics of fractional Thomson oscillator. Physics of Wave Processes and Radio Systems. 2012; **15**(1):64-68

[33] Schroeder M. Fractals, Chaos, Power Laws: Minutes From an Infinite Paradise. New York: V.H. Freeman; 1991

[34] Gerasimov A. Generalization of linear deformation laws and their application to internal friction

problems. AS USSR. Applied Mathematics and Mechanics. 1948;**12**: 529-539

[35] Caputo M. Linear models of dissipation whose Q is almost frequency independent-II. Geophysical Journal International. 1967;**13**(5):529-539

[36] Parovik R. Finite-difference schemes for fractal oscillator with a variable fractional order. Bulletin KRASEC. Physical and Mathematical Sciences. 2015;**11**(2):85-92. DOI: 10.18454/2313-0156-2015-11-2-85-92

[37] Parovik R. Explicit finite-difference scheme for the numerical solution of the model equation of nonlinear hereditary oscillator with variable-order fractional derivatives. Archives of Control Sciences. 2016;**26**(3):429-435. DOI: 10.1515/acsc-2016-0023

[38] Xu Y, Erturk V. A finite difference technique for solving variable-order fractional integro-differential equations. Iranian Mathematical Society. 2014; **40**(3):699-712

[39] Parovik R. Mathematical Modeling of Hereditary Linear Oscillators. Petropavlovsk-Kamchatsky: Vitus Bering Kamchatka State University; 2015. 178 p

[40] Parovik R. Fractional calculus in the theory of oscillating systems. Modern Science Technologies. 2017;**1**:61-68

[41] Daub EG, Carlson JM. Stick-slip instabilities and shear strain localization in amorphous materials. Physical Review E. 2009;**80**(6):066113

[42] Rekhviashvili S. Sh. Dimensional phenomena in condensed matter physics and nanotechnology. Nalchik, Russia: KBNts RAS; 2014

[43] Scholz Ch H. The mechanics of earthquakes and faulting. Cambridge: Cambridge university press; 2002

[44] Xiao-Jun Yang, Tenreiro Machado JA. A new fractional operator of variable order: Application in the description of anomalous diffusion. Physica A: Statistical Mechanics and its Applications. 2017;**481**:276-283

[45] Gao Feng, Xiao-Jun Yang. A new family of the local fractional PDEs. Fundamenta Informaticae. 2017; **151**(1-4):63-75

[46] Boyadzhiev D, Hristo Kiskinov H, Veselinova M, Zahariev A. Stability analysis of linear distributed order fractional systems with distributed delays. Fractional Calculus and Applied Analysis. 2017;**20**(4):914-935

[47] Gallegos JA, Manuel A. Robustness and convergence of fractional systems and their applications to adaptive schemes. Fractional Calculus and Applied Analysis. 2017;**20**(4):895-913

[48] Xiao-Jun Yang, Tenreiro Machado JA, Cattani Carlo, Gao Feng. On a fractal LC-electric circuit modeled by local fractional calculus. Communications in Nonlinear Science and Numerical Simulation. 2017;**47**:200-206

Chapter 2

Quantum Harmonic Oscillator

Coşkun Deniz

Abstract

Quantum harmonic oscillator (QHO) involves square law potential (x^2) in the Schrodinger equation and is a fundamental problem in quantum mechanics. It can be solved by various conventional methods such as (i) analytical methods where Hermite polynomials are involved, (ii) algebraic methods where ladder operators are involved, and (iii) approximation methods where perturbation, variational, semiclassical, etc. techniques are involved. Here we present the general outcomes of the two conventional semiclassical approximation methods: the JWKB method (named after Jeffreys, Wentzel, Kramers, and Brillouin) and the MAF method (abbreviated for "modified Airy functions") to solve the QHO in a very good precision. Although JWKB is an approximation method, it interestingly gives the exact solution for the QHO except for the classical turning points (CTPs) where it diverges as typical to the JWKB. As the MAF method, it enables very approximate wave functions to be written in terms of Airy functions without any discontinuity in the entire domain, though, it needs careful treatment since Airy functions exhibit too much oscillatory behavior. Here, we make use of the parity conditions of the QHO to find the exact JWKB and approximate MAF solutions of the QHO within the capability of these methods.

Keywords: Schrodinger equation, quantum mechanics, JWKB, MAF

1. Introduction

Time-independent Schrodinger equation (TISE) is an eigenvalue problem in the form:

$$\hat{H}|\varphi\rangle = E|\varphi\rangle \Rightarrow \left[\frac{-\hbar^2}{2m}\nabla^2 + U(r)\right]\varphi_n = E_n\varphi \tag{1}$$

where the terms are in the usual meanings, namely, ∇^2, the Laplacian operator; $\hat{H}$, Hamiltonian operator (kinetic energy plus potential energy operators); m, mass; $\hbar$, Planck's constant divided by 2π; φ, wave function (eigenfunction); E, total energy (eigenvalue); and $U(r)$, function of potential energy [1–7]. Quantum harmonic oscillator (QHO) is described by the TISE in (1) for the square law potential:

$$U(r) = \frac{1}{2}mw^2r^2 = \frac{kr^2}{2m} \geq 0 \tag{2}$$

where $w = \sqrt{k/m}$ is the natural angular momentum (associated with the angular frequency $f = w/(2\pi)$). Since $U(r) \geq 0$, our eigenvalue problem (or bound-state problem) requires $E_n \geq 0$ to give the following:

$$\nabla^2 \varphi_n(E_n \geq 0, r) + f(E_n \geq 0, r)\varphi_n(E_n \geq 0, r) = 0;$$
$$f(E_n \geq 0, r) = k^2(E_n \geq 0, r) = \frac{2m}{\hbar^2}[E_n - U(r)] = \frac{2m}{\hbar^2}\left[E_n - \frac{1}{2}mw^2r^2\right] =: \begin{cases} 0 \; or \; (+) \\ (-) \end{cases} \quad (3)$$

The QHO is a very good approximation in solving systems of diatomic molecules vibrating under the spring constant [1, 2, 5] and finds various modern physics applications such as in [8–10] as stated in a famous quotation: "the career of a young theoretical physicist consists of treating the harmonic oscillator in ever-increasing levels of abstraction by Sidney Coleman" [10, 11]. Here, $U(r)$ is central potential which can be given in Cartesian coordinates (x, y, z) where solutions involve Hermite polynomials as in [1–7, 12] or in spherical coordinates (r, θ, ϕ) where solutions involve spherical harmonics as in [1, 2, 13]. For simplicity, it is widely studied in one dimension (say, in x only), and higher dimensional systems are called isotropic harmonic oscillators in 2D or in 3D. The QHO can be solved by various conventional methods such as the following: (i) by analytical methods where some analytic functions involving Hermite polynomials are involved [1–5]; (ii) algebraic methods where ladder operators are involved, that is, [1, 2]; and (iii) by approximation methods such as perturbation methods, JWKB method, variational methods, etc., that is, [1–6, 14]. Brownian study of QHO as an open dynamic quantum system in terms of quantum Langevin equation was studied in [15–17]. We study here one dimensional and non-frictional, that is, undamped case, and present its solution by the two following conventional semiclassical approximation methods: (i) the JWKB method (named after the authors, Jeffreys, Wentzel, Kramers, and Brillouin, who contributed to the theory) [1–7, 14] and (ii) the MAF method (abbreviated from modified Airy function) [3, 18–23].

JWKB method is known to give exact eigenenergies for the QHO, but eigenfunctions fail at and around the classical turning points (CTPs) where $f = 0$ (or, equivalently, $E_n = U(r)$) in (3) as typical to the JWKB method [1–7, 14]. These discontinuities prevent us from using continuity at the boundaries by equating the JWKB solutions of two neighboring regions directly at the CTPs to find the eigenenergy-dependent coefficients in the general JWKB eigenfunctions (wave functions). It also prohibits the use of normalizability of the eigenfunctions between $-\infty$ and ∞. To surmount the problem, parity conditions of the problem regarding the symmetry of the QHO in the dimensionless form are used, and advanced computational software such as Mathematica can be used to achieve these calculations [3, 4, 14, 24]. Moreover, asymptotic matching is required in the JWKB solutions to maintain the normalizability except for the CTPs as discussed above [4, 7, 14]. As to the MAF method, it does not exhibit discontinuities at the CTPs, though highly oscillating behavior of the Airy functions requires careful handling in finding their zeros and the parity treatment used in the JWKB solution seems straightforward to be also applicable to the MAF solution of the QHO [3, 19–22]. Although it was originally suggested in 1931 by Langer in [25], finding zeros of highly oscillatory Airy functions became practical as the advances in computational software and the MAF method became widespread by the 1990s [3, 18–23]. In this work, we present the general outcomes of the conventional JWKB and MAF methods as two semiclassical conventional methods and solve the QHO by using the parity condition of the problem in the dimensionless form pedagogically. We will discuss the treatment of parity matching and asymptotic matching in solving the QHO by these semiclassical methods.

2. Exact solution of the QHO in 1D by the analytic method

The QHO in (3) is a bound-state problem which can be written in 1D for the potential function in 1D ($U(x) = \frac{1}{2}mw^2x^2 = \frac{kx^2}{2m} \geq 0$) as follows:

$$\varphi_n''(x) + f(E_n \geq 0, x)\varphi_n(x) = 0$$

$$f(E_n \geq 0, x) = k^2(E_n \geq 0, x) = \frac{2m}{\hbar^2}[E_n - U(x)] = \frac{2m}{\hbar^2}\left[E_n - \frac{1}{2}mw^2x^2)\right] =: \begin{cases} 0 \; or \; (+) \\ (-) \end{cases} \quad (4)$$

or, simply,

$$\varphi_n''(x) + \frac{2m}{\hbar^2}\left(E_n - \frac{1}{2}mw^2x^2\right)\varphi_n(x) = 0 \quad (5)$$

whose solution by various conventional approaches (such as analytical, algebraic, approximation, etc.) is given in any fundamental textbooks, that is, [1–3, 5] and whose results can be summarized as follows [14]:

i. Change of variable in (4) and (5):

$$y(x) = \beta x = \sqrt{\frac{mw}{\hbar}}x \quad (6)$$

ii. TISE for the QHO in 1D in dimensionless form:

$$\varphi_n''(y) + k^2[\lambda_n(E_n \geq 0), y]\varphi_n(y) = 0$$

$$\Leftrightarrow \left\{ f[\lambda_n(E_n) \geq 0, y] =: k^2[\lambda_n(E_n) \geq 0, y] = \lambda_n^2 - y^2, \lambda_n^2 = \frac{2E_n}{\hbar w} =: \left\{ \begin{matrix} 0 \; or \; (+) \\ (-) \end{matrix} \right\} \right. \quad (7)$$

Note that here $f =: k^2$ is a function of $\lambda_n(E_n)\&y$ and $\lambda_n \geq 0$ since $E_n \geq 0$. Moreover, $f(\lambda_n, y)$ is an even function as shown in **Figure 1** (λ is chosen as continuous including the discrete energy values assuming that the eigenenergies have not been found yet).

iii. Exact eigenenergies:

$$\Lambda_{EX} =: \Lambda = \lambda_n^2 = 2n + 1 \Rightarrow \boxed{E_n = \left(n + \frac{1}{2}\right)\hbar w, n = 0, 1, 2, \ldots} \quad (8)$$

iv. Exact eigenfunctions (wave functions) in y:

$$\varphi_n(y) =: \psi(\lambda_n, y) = \sqrt{\frac{1}{\sqrt{\pi}2^n n!}} H_n(y)^{-\frac{y^2}{2}}, n = 0, 1, 2, \ldots \quad (9)$$

v. By using (6), we have the wave functions in x:

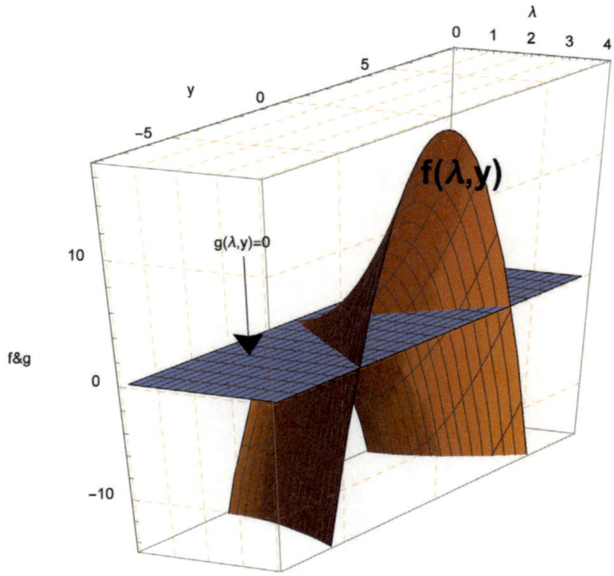

Figure 1.
Graphs of $f(\lambda \geq 0, y)$ *and* $g(\lambda \geq 0, y) = 0$.

$$\varphi_n(x) =: \varphi(\beta, \lambda_n, x) = \sqrt{\frac{\beta}{\sqrt{\pi}2^n n!}} H_n(\beta x)^{-\frac{(\beta x)^2}{2}}, n = 0, 1, 2, \ldots \qquad (10)$$

We used two different symbols (φ and ψ) to label the functions in two different independent variables (in x and in y, respectively). Exact wave functions in (9) (via exact eigen energies in (8)) are given for even and odd n values in **Figures 3** and **4** along with the JWKB solutions for comparison. H_n in (9) and (10) is Hermite polynomials with indice n (named after the French mathematician Charles

Rodriguez formula: $H_n(x) = (-1)^n e^{(x^2)} \frac{d^n}{dx^n} e^{(-x^2)}$
Generating function: $\exp(2xt - t^2) = \sum_{n=0}^{\infty} \frac{H_n(x)t^n}{n!}$
Some of the Hermite polynomials: $H_0(x) = 1, H_1(x) = 2x, H_2(x) = 4x^2 - 1$ $H_3(x) = 8x^3 - 12x, H_4(x) = 16x^4 - 48x^2 + 12$
Recurrence relations: $H_{n+1} = 2xH_n(x) - 2nH_{n-1}(x)$ $H'_n(x) = 2nH_{n-1}(x)$
Evenness and oddity $H_n(-x) = (-1)^n H_n(x)$ $\therefore H_n(x) = \begin{cases} \text{odd, if } n \text{ is odd} \\ \text{even, if } n \text{ is even} \end{cases}$
Orthogonality: $\int_{-\infty}^{\infty} e^{-x^2} H_m(x) H_n(x) dx = \begin{cases} 0 \text{ if } m \neq n \\ 2^n n! \sqrt{\pi} \text{ if } m = n \end{cases}$

Table 1.
Some properties of Hermite polynomials.

Hermite). Some of the properties of Hermite polynomials are tabulated in **Table 1**, and calculation of conversion factor β which exhibits a quantization, namely,

$$\beta = \sqrt{\frac{mw}{\hbar}}\bigg|_{w=\sqrt{\frac{k}{m}}} = \left(\frac{mk}{\hbar^2}\right)^{1/4} = \frac{k}{\hbar w} = \sqrt{\frac{(n+1/2)k}{E_n}} =: \beta_n, \tag{11}$$

is given along with the related Mathematica codes in [14].

3. A review of the JWKB solution of the QHO

2D plot of **Figure 1** is schematically given in **Figure 2** for the QHO under study (in the dimensionless form) from which we have the following outcomes [14]:

3.1 JWKB eigenenergies of the QHO

JWKB eigenenergies can be found by applying the Bohr-Sommerfeld quantization formula given by [1–7, 14]:

$$\int_{y_1}^{y_2} k\left[\tilde{\lambda}_J(E_n), y\right]dy = \left(n + \frac{1}{2}\right)\pi \tag{12}$$

as follows:

$$\begin{aligned}
&\int_{y1=-\lambda_n}^{y_2=\lambda_n} k\left[\tilde{\lambda}_J(E_n), y\right]dy = 2\int_0^{y_2=\lambda_n} k\left[\tilde{\lambda}_J(E_n), y\right]dy = \left(n+\frac{1}{2}\right)\pi \\
&\Rightarrow \int_{-\lambda_n}^{\lambda_n} \sqrt{\tilde{\lambda}_J^2(E_n) - y^2 dy} = 2\int_0^{\lambda_n} k\left[\tilde{\lambda}_J(E_n), y\right]dy = \left(n+\frac{1}{2}\right)\pi \\
&y =: \sin\theta \Rightarrow 2\textstyle\int_{-\pi/2}^{\pi/2} \tilde{\lambda}_J^2(E_n)\cos^2\theta d\theta = \left(n+\frac{1}{2}\right)\pi \Leftrightarrow \tilde{\lambda}_J^2(E_n)\frac{\pi}{2} = \left(n+\frac{1}{2}\right)\pi \\
&\Rightarrow \boxed{\tilde{\lambda}_J(E_n) = \sqrt{2n+1}\,(\text{or})\,\tilde{E}_{Jn} =: E_{JWKB,n} = \left(n+\frac{1}{2}\right)\hbar\omega}
\end{aligned} \tag{13}$$

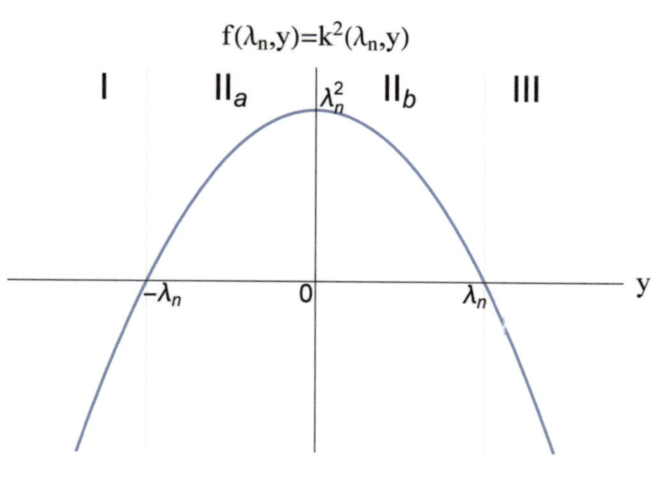

Figure 2.
Schematic 2D sketch of $f[\lambda_n(E_n) \geq 0, y] =: k^2[\lambda_n(E_n \geq 0), y]$ *for a given* λ_n.

which is already the exact solution given in (8) [1–4, 6, 7, 14]. Results are given along with the MAF solutions in **Table 2** for comparison. Note that we use the following notation for the symmetrical (or even parity (EP)) and antisymmetrical (or odd parity (OP)) solutions:

$$\tilde{E}_{Jn} = \begin{cases} \tilde{E}_{J,n_s} = (n_s + 1/2)\pi, n_s = 0, 2, 4, \ldots \\ \tilde{E}_{J,n_a} = (n_a + 1/2)\pi, n_a = 1, 3, 5, \ldots \end{cases} \quad (14)$$

where the subscripts "J, n_s" represent J, JWKB, and n_s, symmetrical indices (n_s = even), and similarly, "J, n_a" represents J, JWKB, and n_a, antisymmetrical indices (n_s =odd).

3.2 JWKB solution of eigenfunctions (wave functions) of the QHO

Conventional first-order JWKB solution of the QHO given in the normal form in (4) or (7) is as follows:

$$\boxed{\tilde{\varphi}_J(\lambda_n, y) = \frac{c_{J1} \exp\left[-i \int^{y_t, y} k(\lambda_n, y) dy\right]}{\sqrt{k(\lambda_n, y)}} + \frac{c_{J2} \exp\left[i \int^{y_t, y} k(\lambda_n, y) dy\right]}{\sqrt{k(\lambda_n, y)}}} \quad (15)$$

where y_t is either of the classical turning points (CTPs: either "y_1, on the left" or "y_2, on the right" depending on the region under question) and $c_{J1}\&c_{J2}$ are arbitrary JWKB constants. Once solution in any region is found (say, φ_{JII}), solution in the adjacent region (say, φ_{JIII}) can directly be found by the conventional JWKB connection formulas given in [3, 4, 14] without calculating it via (15). The integrals here are the definite integrals whose upper and lower values should be chosen as the related turning point (either y_1 or y_2) and the variable y should be in the correct ascending integration order. Normally, constant coefficients in the general solutions are determined from normalization by applying the boundary conditions of the

Index (=MAF index)	EP					OP				
	MAF			JWKB*		MAF			JWKB*	
$n = n_M$	Z_{n_s}	ζ_{n_s}	$\tilde{E}_{M,n_s}$	n_a	$\tilde{E}_{J,n_s}$	Z_{n_a}	ζ_{n_a}	$\tilde{E}_{M,n_a}$	n_a	$\tilde{E}_{J,n_a}$
0	1.20348	0.0603317	0.5603317ℏω	0	0.5ℏω	2.33811	1.01735	1.51735ℏω	1	1.5ℏω
1	3.27162	2.0115	2.5115ℏω	2	2.5ℏω	4.08795	3.0079	3.5079ℏω	3	3.5ℏω
2	4.83082	4.0063	4.5063ℏω	4	4.5ℏω	5.52056	5.00508	5.50508ℏω	5	5.5ℏω
3	6.16988	6.00435	6.50435ℏω	6	6.5ℏω	6.78671	7.00374	7.50374ℏω	7	7.5ℏω
4	7.37677	8.00332	8.50332 ℏω	8	8.5ℏω	7.94413	9.00295	9.50295ℏω	9	9.5ℏω
5	8.49195	10.0027	10.5027ℏω	10	10.5ℏω	9.02265	11.0024	11.5024ℏω	11	11.5 ℏω
6	9.5382	12.0023	12.5023ℏω	12	12.5ℏω	10.0402	13.0021	13.5021ℏω	13	13.5ℏω
7	10.5299	14.0019	14.5019ℏω	14	14.5ℏω	11.0085	15.0018	15.5018ℏω	15	15.5ℏω

**JWKB solution is exact.*

Table 2.
JWKB and MAF eigenenergies.

eigenvalue problem. However, it is useless since the boundary conditions correspond to the CTPs at which (and also in a narrow region around) the conventional first-order JWKB solutions typically diverge [3, 4, 7, 14]. This might be thought as a violation of continuity requirement of the acceptable wave function properties concerning continuity, but higher order JWKB approximation can fix it. Now, due to the discontinuities at the boundaries between the adjacent regions (such as between I and II or between II and III), the unidirectional JWKB connection formulas (surely, for the first-order JWKB) given in the literature [3, 4, 7, 14] cannot be used to find the constant coefficients in the general solution in (18) and (19). Note that these connection formulas can be used to determine the structure of the JWKB solutions in all regions (I, II, and III), but they cannot be used to find the constant coefficients (which will be a function of eigenenergy) as explained.

However, we are fortunately not helpless: since $f[\lambda_n(E_n)\geq 0, y] =: k^2[\lambda_n(E_n\geq 0), y]$ in (7) is an even function (see **Figure 2**), we should have even and odd-parity solutions. If we start by considering the exact solutions in (9) and (10) and considering them to be approximate to the JWKB solution (shown with tilde and subscript J), we have the following outcomes [14]:

$$\boxed{\psi(\lambda_n, y) = [\varphi(\beta, \lambda_n, x\rightarrow y/\beta)]_{\beta=1}\text{OR} : \varphi(\beta, \lambda_n, x) = \sqrt{\beta}\psi(\lambda_n, y\rightarrow \beta x)} \tag{16}$$

$$\begin{cases} \text{E.P.} : \varphi_n(\beta, \lambda, -x) = \varphi_n(\beta, \lambda, x), n = 0, 2, 4, \ldots \Rightarrow \begin{cases} i)\, \varphi_n(\beta, \lambda, 0) = \pm p, p > 0 \\ ii)\, \left.\frac{\partial \varphi_n(\beta, \lambda, x)}{\partial x}\right|_{x=0} = 0 \end{cases} \\ \text{O.P.} : \varphi_n(\beta, \lambda, -x) = -\varphi_n(\beta, \lambda, x), n = 1, 3, 5, \ldots \Rightarrow \begin{cases} i)\, \lambda_n(\beta, \lambda, 0) = 0 \\ ii)\, \left.\frac{\partial \varphi_n(\beta, \lambda, x)}{\partial x}\right|_{x=0} = \pm q, q > 0 \end{cases} \end{cases}$$

$$\Rightarrow \boxed{\beta \simeq \tilde{\beta}_J = \begin{cases} \tilde{\beta} \simeq \beta_{J.E.P.} = \left[\dfrac{\pm p}{\varphi_n(1, \lambda, y)}\right]^2_{y=0}, n = 0, 2, 4, \ldots \\ \tilde{\beta}_{O.P.} \simeq \beta_{J.O.P.} = \left[\dfrac{\pm q}{\dfrac{\partial \varphi_n(1, \lambda, y)}{\partial y}}\right]^2_{y=0}, n = 1, 3, 5, \ldots \end{cases}} \tag{17}$$

where $p\&q$ are positive real constants regarding the even-parity (EP) and odd-parity (OP) initial values of the physical system. Remember that we use φ for the x-system and ψ for the y-system as shown in (16). In finding the constant coefficients, we can take $\pm p = \pm q = 1$, and alternating sign can be modified as a parity matching as follows [14]:

$$\boxed{\tilde{\psi}^{(\text{par.m.})}_{(J,E.P.)}(\lambda_n, y) = (-1)^{\left(\frac{n}{2}\right)} \times \left\{ \begin{array}{l} \tilde{\psi}^{(\text{asy.m.})}_{J,I}(\lambda_n, y) = \tilde{\psi}^{(\text{asy.m.})}_{J,III}(\lambda_n, -y) \text{ for } -\infty < y \leq -\lambda_n \\ \tilde{\psi}_{J,II}(\lambda_n, y) = \begin{cases} \tilde{\psi}_{J,II}(\lambda_n, -y) \text{ for } -\lambda_n < y \leq 0 \\ \tilde{\psi}_{J,II}(\lambda_n, y) \text{ for } 0 < y \leq \lambda_n \end{cases} \\ \tilde{\psi}^{(\text{asy.m.})}_{J,III}(\lambda_n, y) \text{ for } \lambda_n \leq y < \infty \end{array} \right\}_{p=1}} \tag{18}$$

$$\tilde{\psi}^{(\text{par.m.})}_{(J,O.P.)}(\lambda_n, y) = (-1)^{\left(\frac{n-1}{2}\right)} \times \left\{ \begin{array}{l} \tilde{\psi}^{(\text{asy.m.})}_{J,I}(\lambda_n, y) = -\tilde{\psi}^{(\text{asy.m.})}_{J,III}(\lambda_n, -y) \text{ for } -\infty < y \le -\lambda_n \\ \tilde{\psi}_{J,II}(\lambda_n, y) = \left\{ \begin{array}{l} -\tilde{\psi}_{J,II}(\lambda_n, -y) \text{ for } -\lambda_n < y \le 0 \\ \tilde{\psi}_{J,II}(\lambda_n, y) \text{ for } 0 < y \le \lambda_n \end{array} \right. \\ \tilde{\psi}^{(\text{asy.m.})}_{J,III}(\lambda_n, y) \text{ for } \lambda_n \le y < \infty \end{array} \right\}_{q=1} \tag{19}$$

where the superscripts (*par.m.*) and (*asy.m.*) represent *parity matched* and *asymptotically matched* JWKB solutions, respectively. Eqs. (18) and (19) tells that we will take $\pm p = \pm q \to 1$ to find the solutions in $\{0 \le y \le y_2 = \lambda_n\} \cup \{y_2 = \lambda_n \le y < \infty\}$ firstly by using (21) for the asymptotic matching and then extending it to the second quadrant according to the parity under question. Note that asymptotically matched general $(JWKB)_1$ solution can be obtained as follows (see [3, 4, 7, 14] for details):

$$\tilde{\varphi}^{(asy.m.)}_{J}(\lambda_n, y) = \left\{ \begin{array}{l} \tilde{\varphi}^{(asy.m.)}_{JI}(\lambda_n, y) = \text{either } \tilde{k}_1 \tilde{\varphi}_1(\lambda_n, y < y_{t1}) \text{ or } \tilde{k}_2 \tilde{\varphi}_2(\lambda_n, y < y_{t1}) \\ \tilde{\varphi}_{JII}(\lambda_n, y) = \tilde{\varphi}_J(\lambda_n, y_{t1} < y < y_{t2}) \\ \tilde{\varphi}^{(asy.m.)}_{JIII}(\lambda_n, y) = \text{either } \tilde{k}_1 \tilde{\varphi}_{J1}(\lambda_n, y_{t2} < y) \text{ or } \tilde{k}_2 \tilde{\varphi}_{J2}(\lambda_n, y_{t2} < y) \end{array} \right. \tag{20}$$

so that they exhibit the following asymptotic behaviors:

$$\tilde{\varphi}^{(asy.m).}_{J}(\lambda_n, y) = \left\{ \begin{array}{l} \lim\limits_{y \to -\infty} \left[\tilde{\varphi}_J\left(\lambda_n, y < y_{t_1}\right) = \tilde{\varphi}_{JI}(\lambda_n, y) \right] = 0 \\ \lim\limits_{y \to \infty} \left[\tilde{\varphi}_J(\lambda_n, y_{t2} < y) = \tilde{\varphi}_{JIII}(\lambda_n, y) \right] = 0 \end{array} \right. \tag{21}$$

3.2.1 *Even-parity (EP) wave functions*

When initial values at $x = y = 0$ for the EP case in (17), namely (by using (16)),

$$\left\{ \tilde{\psi}_{J,II}(\lambda_n, y)\Big|_{y=0} = 1/\sqrt{\beta}; \partial_y \tilde{\psi}_{J,II}(\lambda_n, y)\big|_{y=0} = 0 \right\}, \tag{22}$$

is applied to the JWKB solution in (15), we find the following:

$$i)\ \partial_y \tilde{\psi}_{J,II}(\lambda_n, y)\big|_{y=0} = 0 \Rightarrow \tilde{\psi}_{II}(\tilde{\lambda}_n = \lambda_n, y) = \frac{A(\lambda_n)}{\sqrt{k(\lambda_n, y)}} \cos\left[\int_0^y k(\lambda_n, y) dy \right], \text{ for } 0 < y < \lambda_n \tag{23}$$

where the second complementary solution (in the sine form) has been canceled and calculation of the integral in the cosine term can be calculated by the similar change of variable as in (13) whose result will give $\eta(y, 0)$ (see Eq. (18) below and apply $\eta(y, 0) = \eta(\lambda_n \to y, y \to 0)$).

$$ii)\ \tilde{\psi}_{J,II}(\lambda_n, y)\big|_{y=0} = 1/\sqrt{\beta} \Rightarrow A(\lambda_n) = \sqrt{\frac{\lambda_n}{\beta}} \tag{24}$$

and by using (16), we have.

$$\tilde{\psi}_{J,II}(\beta,\lambda_n,x) = \sqrt{\beta}\tilde{\psi}_{J,II}(\lambda_n,y\to\beta x) = \sqrt{\frac{\lambda_n}{k(\lambda_n,\beta x)}}\cos\left[\int_0^{\beta x} k(\lambda_n,y)dy\right], \text{for } 0<x<\lambda_n/\beta \tag{25}$$

Now, by applying the JWKB connection formula with a small phase term α, we get.

$$\tilde{\psi}_{III}(\tilde{\lambda}_n=\lambda_n,y) = \frac{A(\lambda_n)}{\sqrt{\kappa(\lambda_n,y)}}\left\{\begin{array}{l}\cos[\alpha(\lambda_n)]\exp[\zeta(\lambda_n,y)] \\ +\frac{1}{2}\sin[\alpha(\lambda_n)]\exp[-\zeta(\lambda_n,y)]\end{array}\right\}, \text{for } \lambda_n<y<\infty, \tag{26}$$

and the asymptotically matched (modified) wave function in region III via (20) and (21) of [3, 4, 7, 14] gives:

$$\tilde{\psi}_{III}^{(asy.m.)}(\tilde{\lambda}_n=\lambda_n,y) = \frac{A(\lambda_n)}{2\sqrt{\kappa(\lambda_n,y)}}\sin[\alpha(\lambda_n)]\exp[-\zeta(\lambda_n,y)], \text{ for } \lambda_n<y<\infty \tag{27}$$

Abbreviations we use for the EP JWKB solutions here (and also for the OP solutions in the next subsection) are as follows [14]:

$$\left\{\begin{array}{l} \alpha(\lambda_n) = \int_0^{\lambda_n} k(\lambda_n,y)dy + \frac{\pi}{4} = \eta(\lambda_n,0)+\frac{\pi}{4} \\ \eta(\lambda_n,y) = \int_y^{\lambda_n} k(\lambda_n,y)dy \to = \frac{\lambda_n^2\pi}{4} - \frac{\lambda_n^2}{2}\sin^{-1}\left(\frac{y}{\lambda_n}\right) - \frac{y}{2}\sqrt{\lambda_n^2-y^2} \\ \zeta(\lambda_n,y) = \int_{\lambda_n}^{y} \kappa(\lambda_n,y)dy \to = \frac{y}{2}\sqrt{y^2-\lambda_n^2} - \frac{1}{2}\lambda_n^2\ln\left|\frac{y+\sqrt{y^2-\lambda_n^2}}{\lambda_n}\right| \end{array}\right. \tag{28}$$

Since we have already calculated $\tilde{\psi}_{J,II}(\beta,\lambda_n,x)$ and $\tilde{\psi}_{III}^{(asy.m.)}(\tilde{\lambda}_n=\lambda_n,y)$ in the first quadrant $(0\le y\le\lambda_n)$, JWKB solutions in the other regions can be easily written as in (18). JWKB wave functions regarding the EP case are given in **Figure 3** along with the exact solutions for comparison.

3.2.2 Odd-parity (OP) wave functions

Similarly, by using the boundary conditions for the OP case in (17), namely (by using (16)),

$$\left\{\tilde{\psi}_{J,II}(\lambda_n,y)\Big|_{y=0} = 0; \partial_y\tilde{\psi}_{J,II}(\lambda_n,y)\big|_{y=0} = 1/\sqrt{\beta^3}\right\}, \tag{29}$$

and starting with region II.

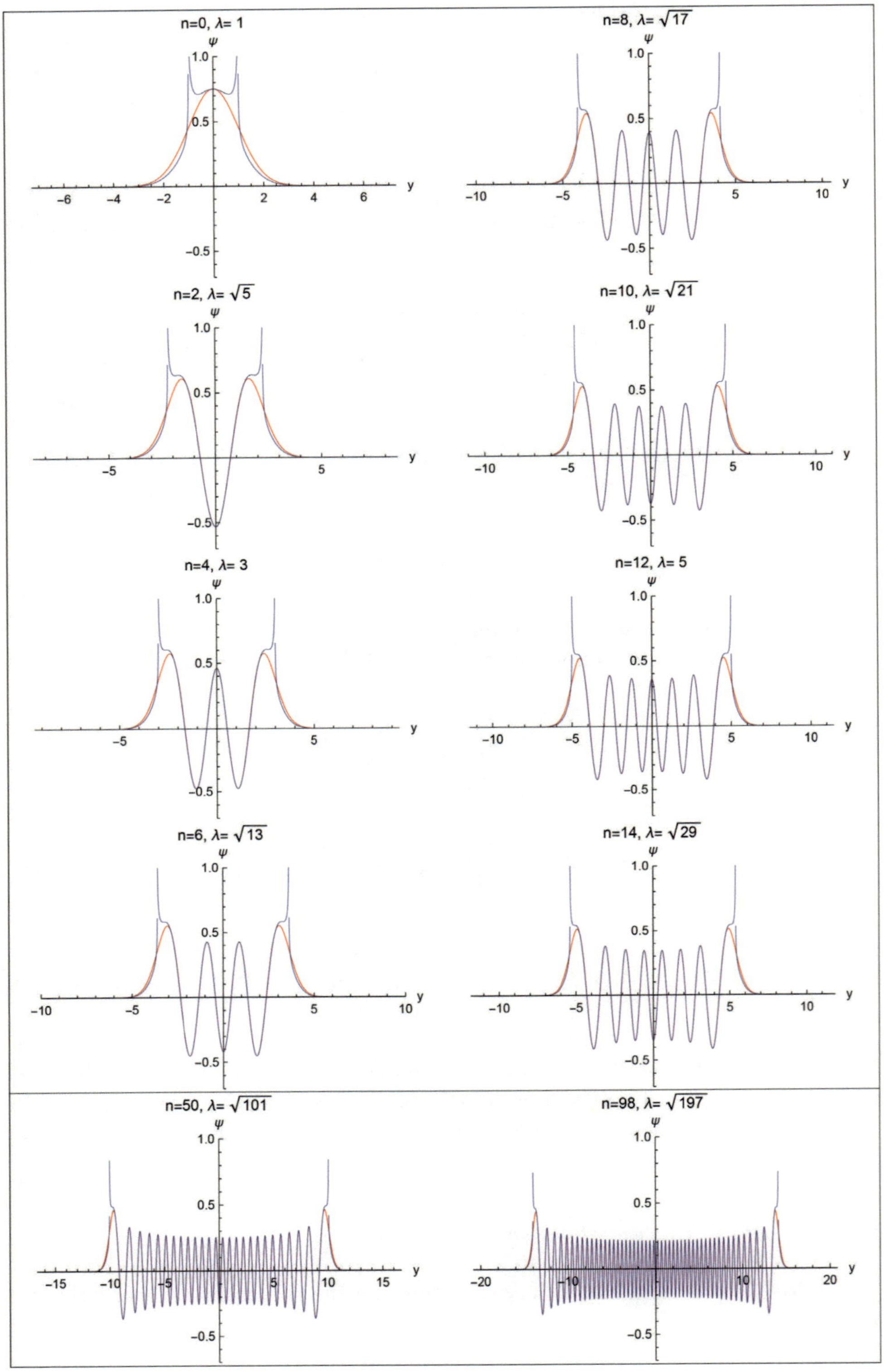

Figure 3.
Exact and JWKB solutions of EP wave functions (for $p = 1$).

$$\text{(i)}\quad \tilde{\psi}_{J,II}(\lambda_n,y)\big|_{y=0}=0\Rightarrow\tilde{\psi}_{II}(\tilde{\lambda}_n=\lambda_n,y)=\frac{B(\lambda_n)}{\sqrt{k(\lambda_n,y)}}\sin\left[\int_0^y k(\lambda_n,y)dy\right],\ \text{for}\ 0<y<\lambda_n \tag{30}$$

$$\text{(ii)}\quad \partial_y\tilde{\psi}_{J,II}(\lambda_n,y)\big|_{y=0}=1/\sqrt{\beta^3}\Rightarrow B(\lambda_n)=1/\sqrt{\lambda_n\beta^3} \tag{31}$$

connecting to region III in the first quadrant ($0\le y\le\lambda_n$) via the JWKB connection formula.

$$\tilde{\psi}_{J,III}(\tilde{\lambda}_n=\lambda_n,y)=\frac{B(\lambda_n)}{\sqrt{\kappa(\lambda_n,y)}}\left\{\begin{array}{c}\sin[\alpha(\lambda_n)]\exp[\zeta(\lambda_n,y)]\\ -\frac{1}{2}\cos[\alpha(\lambda_n)]\exp[-\zeta(\lambda_n,y)]\end{array}\right\},\ \text{for}\ \lambda_n<y<\infty \tag{32}$$

whose asymptotic matching gives.

$$\tilde{\psi}^{(m.)}_{J,III}(\lambda_n,y)=-\frac{B(\lambda_n)}{2\sqrt{\kappa(\lambda_n,y)}}\cos[\alpha(\lambda_n)]\exp[-\zeta(\lambda_n,y)],\ \text{for}\ \lambda_n<y<\infty \tag{33}$$

Again, since we have already obtained $\tilde{\psi}_{J,II}(\lambda_n,y)$ and $\tilde{\psi}^{(m.)}_{J,III}(\lambda_n,y)$ in $0\le y\le\lambda_n$, JWKB solutions in the second quadrant can be written in terms of them as shown in (19). JWKB wave functions regarding the OP case are given in **Figure 3** along with the exact solutions for comparison.

4. The MAF method

If we follow the QHO in dimensionless form given in (7), we have the following properties in MAF theories [3, 18–23]:

4.1 General structure of the MAF approximation to the bound-state wave functions

Formal MAF method suggests a solution to the TISE in (7) in terms of Airy functions as follows:

$$\begin{array}{l}\psi_{MAF}(\lambda_n,y)=:\tilde{\psi}_M(\lambda_n,y)=\left\{\begin{array}{c}F(\lambda_n,y)Ai[\xi(\lambda_n,y)]\\ or\\ G(\lambda_n,y)Bi[\xi(\lambda_n,y)]\end{array}\right\}\\ \Rightarrow\tilde{\psi}_M(\lambda_n,y)=a_1F(\lambda_n,y)Ai[\xi(\lambda_n,y)]+a_2G(\lambda_n,y)Bi[\xi(\lambda_n,y)]\end{array} \tag{34}$$

where Ai and Bi represent the Airy functions (namely, $Ai(x)$ and $Bi(x)$ are the linearly independent solutions of the Airy differential equation $y''(x)-xy(x)=0$ in x), $a_1\&a_2$ are the arbitrary constants which will be found from boundary values, and $F\&G$ are the functions to be determined. Note that the first variable λ_n is the eigenenergies (constant values quantized by index n) which will also be determined soon. So, for now, we can consider all these functions as one dimensional in only y for simplification (say, $\tilde{\psi}_J(\lambda_n, y) =: \tilde{\psi}_J(y)$, $\xi(\lambda_n, y) =: \xi(y)$, $F(\lambda_n, y) =: F(y)$, etc.). If we choose one of the linearly independent solutions, say $F(y).Ai[\xi(y)]$, to substitute in the TISE in (4), then it gives:

$$\frac{F''(y)}{F(y)} + \frac{2F'(y)Ai'[\xi(y)]\xi'(y) + F(y)Ai'[\xi(y)]\xi''(y)}{F(y)Ai[\xi(y)]} + \left\{\xi(y)[\xi'(y)]^2 + f(y)\right\} = 0 \quad (35)$$

Now, with the choice of the last term in (35) as zero, we find the following:

$$\xi(y)[\xi'(y)]^2 + f(y) = 0 \Rightarrow \xi(y) = \left[\int_{y_t}^{y} \frac{3}{2}\sqrt{-f(y)dy}\right]^{2/3} \quad (36)$$

Here, the property of the Airy functions, $Ai''(\xi) = \xi Ai(\xi)$, was used [3, 18]. The integral interval in (36) is also chosen tactically in a fashion that it invokes a relationship with the turning point y_t (representing the correct order y_{t1} or y_{t2} to give a non-imaginary result), and it can be written in a more explicit and conventional form (by also using in our two-variable form here) as follows:

$$\xi(\lambda_n, y) = \begin{cases} \xi_I : \left[\frac{3}{2}\int_y^{y_{t1}} \kappa(\lambda_n, y)dy\right]^{2/3}, for\ y \leq y_{t1} \\ \xi_{II} : -\left[\frac{3}{2}\int_{y_{t1}}^{y} k(\lambda_n, y)dy\right]^{2/3} = -\left[\frac{3}{2}\int_y^{y_{t2}} k(\lambda_n, y)dy\right]^{2/3},\ for\ y_{t1} \leq y \leq y_{t2} \\ \xi_{III} : \left[\frac{3}{2}\int_{y_{t2}}^{y} \kappa(\lambda_n, y)dy\right]^{2/3}, for\ y_{t2} \leq y \end{cases} \quad (37)$$

where $f(\lambda_n, y \geq 0) = k^2(\lambda_n, y) = -\kappa^2(\lambda_n, y)$ and $y_{t1}\&y_{t2}$ are the CTPs at the interface of the regions $I - II\&II - III$, respectively. The remaining terms in (44) and (45) are also made zero as follows:

Starting from the second term, we have.

$$\frac{2F'Ai'(\xi)\xi'}{F} + Ai'(\xi)\xi'' = 0 \Rightarrow F(y) = \frac{b_1}{\sqrt{\xi'(y)}} \quad (38)$$

where b_1 is some constant, and finally, making the first term in (35) zero (which is the only assumption in the MAF method), we have the following:

$$P(y) = \frac{F''(y)}{F(y)} \approx 0 \quad (39)$$

Or more correctly in two-variable form in our eigenvalue system.

$$P(\lambda_n, y) = \frac{\partial_y^2 F(\lambda_n, y)}{F(\lambda_n, y)} \quad (40)$$

can be thought as a measure of the accuracy of the MAF solution, namely, $P(\lambda_n, y) \to 0$, as MAF solution gets more accurate [18].

The same results would also be obtained if we had chosen the other linearly independent solution, $G(y).Bi(y)$, in (34). Consequently, using the results found in

(37) and (38), the general solution suggested in (34) can be written explicitly in the standard form of the MAF formula as follows:

$$\psi_{MAF}(y) = \frac{\tilde{c}_1}{\sqrt{\xi'(y)}} Ai[\xi(y)] + \frac{\tilde{c}_2}{\sqrt{\xi'(y)}} Bi[\xi(y)] \tag{41}$$

or more correctly in two variables here in our study.

$$\boxed{\psi_{MAF}(\lambda_n, y) = \frac{\tilde{c}_1}{\sqrt{\partial_y \xi(\lambda_n, y)}} Ai[\xi(\lambda_n, y)] + \frac{\tilde{c}_2}{\sqrt{\partial_y \xi(\lambda_n, y)}} Bi[\xi(\lambda_n, y)]} \tag{42}$$

where $c_1 = a_1.b_1$ and $c_2 = a_2.b_2$ are the arbitrary constants to be determined from the boundary values as mentioned and $\partial_y \xi(\lambda, y)$ represents the first derivative of ξ with respect to y. Using the result in (38), the approximation term $P(\lambda_n, y)$ in (40) can be rewritten explicitly as follows:

$$P(\lambda_n, y) = \frac{3}{4} \left[\frac{\partial_y^2 \xi(\lambda_n, y)}{\partial_y \xi(\lambda_n, y)} \right]^2 - \frac{\partial_y^3 \xi(\lambda_n, y)}{2 \partial_y \xi(\lambda_n, y)} \tag{43}$$

4.2 MAF solution of eigenenergies

For a symmetrical f as in **Figure 2**, we have even-parity (EP) and odd-parity (OP) MAF wave functions just as in JWKB method, but now it leads to two different MAF quantization formulas with two different MAF universal constants regarding EP and OP solutions as given in [3] and as we study in this section. We again use the symbolism in (9) ($\varphi \Leftrightarrow \psi$) and start with the first quadrant, by applying that $\lim_{y \to \infty} = 0$ requires $c_2 = 0$ in (42), namely,

$$\psi_{MAF,n}(\lambda_n, y) =: \tilde{\psi}_{Mn}(\lambda_n, y) = \frac{\tilde{c}_1}{\sqrt{\partial_y \xi(\lambda_n, y)}} Ai[\xi(\lambda_n, y)], \tag{44}$$

where the denominator can be written in the following form [3]:

$$\frac{1}{|\sqrt{\partial_y \xi(\lambda_n, y)}|} = \frac{|\xi(\lambda_n, y)|^{1/4}}{|k^2(\lambda_n, y)|^{1/4}} \tag{45}$$

i. Even-parity (EP) eigenenergies: if we apply the EP formulas of the exact solution in (17), by using (16), to the MAF wave functions, we have the following:

$$\left\{ \tilde{\psi}_{Mn}(\lambda_n, 0) = \frac{\pm p}{\sqrt{\beta}}, \partial \frac{\tilde{\psi}_{Mn}(\lambda_n, y)}{\partial y} \Big|_{y=0} = 0 \right\} \tag{46}$$

$$\text{(i)} \quad \tilde{\psi}_M(\lambda_n, 0) = \frac{1}{\sqrt{\beta}} \Rightarrow \text{find } \tilde{c}_1 =: \tilde{c}_{1s}$$

$$\text{result} :\Rightarrow \tilde{c}_{1s} = \frac{\xi_0'}{\sqrt{\beta} Ai[\xi(\lambda_n, y)]} \Rightarrow \tilde{\psi}_M(\lambda_n, y) = \sqrt{\frac{\xi_0'}{\beta \partial_y Ai[\xi(\lambda_n, y)]}} \frac{Ai[\xi(\lambda_n, y)]}{Ai(\xi_0)} \tag{47}$$

$$\text{(ii)} \quad \partial \frac{\tilde{\psi}_M(\lambda_n, y)}{\partial y}\Big|_{y=0} = 0 \Rightarrow Ai'(\xi_0) - \frac{Ai(\xi_0)\xi_0''}{2\xi'^2_0} = 0 \tag{48}$$

where we used ξ_0, ξ_0', and ξ_0'' for simplification, namely, $\xi_0 = \xi(\lambda_n, 0)$, $\xi_0' = \partial_y(\lambda_n, y)\big|_{y=0}$, and $\xi_0'' = \partial_{yy}(\lambda_n, y)\big|_{y=0}$, respectively. We assumed here $\varphi(\lambda_n, 0) = 1$, and we will use then parity correction for $\varphi(\lambda_n, 0) = -1$ case just as in the JWKB calculations. If we take the derivative of (36), we get.

$$\xi'^3(\lambda_n, y) + 2\xi(\lambda_n, y)\left[\partial_y\xi(\lambda_n, y)\right]\left[\partial_{yy}\xi(\lambda_n, y)\right] + \partial_y k^2(\lambda_n, y) = 0 \tag{49}$$

where the last term vanishes as $y \to 0$ to give.

$$\xi'^2_0 + 2\xi_0\xi_0'' = 0 \Rightarrow \frac{\xi_0''}{2\xi'^2_0} = -\frac{1}{4\xi_0} \tag{50}$$

whose substitution in (48) gives.

$$\xi_0 Ai'(\xi_0) + \frac{1}{4}Ai(\xi_0) = 0 \tag{51}$$

Now, by the substitution of $\xi_0 \to -Z_{sn}$, we have.

$$-Z_{sn}Ai'(-Zsn) + \frac{Ai(-Z_{sn})}{4} = 0 \tag{52}$$

where the subscripts sn stand for s, symmetrical solution (EP), and n, quantization order (nth quantization), and Z_{sn} is the nth solution of the differential equation in (52) regarding the symmetrical solution. Now, by using the results in (13), we find the MAF quantization formula regarding the symmetrical solution:

$$\int_0^{y_2} k\left[\tilde{\lambda}_{M,sn}, y\right]dy =: \left(\zeta_{sn} + \frac{1}{2}\right)\pi \Rightarrow \zeta_{sn} = \left(\frac{4Z_{sn}^{3/2}}{3\pi} - \frac{1}{2}\right), n = 1, 2, 3, \ldots \tag{53}$$

where ζ_{sn} is the universal MAF constants regarding the symmetrical solution whose values are given in **Table 2** along with the JWKB solutions (which are already exact) for some n values in comparison. Note that we used n_s to represent the symmetrical (EP) MAF indices in **Table 2**.

ii. Odd-parity (OP) eigenenergies: similarly, if we apply the OP formulas of the exact solution in (17), by using (16), to the MAF wave functions, we have the following:

$$\left\{\tilde{\psi}_M(\lambda, 0) = 0, \partial_y\tilde{\psi}_M(\lambda, y)\big|_{y=0} = \frac{\pm q}{\sqrt{\beta^3}}\right\} \tag{54}$$

$$\text{(i)} \quad \tilde{\psi}_M(\lambda, 0) = 0 \Rightarrow Ai(\xi_0 \to -Z_{an}) = Ai(-Z_{an}) = 0, n = 1, 2, 3, \ldots \tag{55}$$

$$\text{(ii)}\quad \partial_y \tilde{\psi}_M(\lambda_{n_a}, y)\big|_{y=0} = \frac{1}{\sqrt{\beta^3}} \Rightarrow Ai'(\xi_0) - \frac{Ai(\xi_0)\xi_0''}{2\xi'^2_0} = \frac{1}{\sqrt{\beta^3}} \Rightarrow \text{find}\,\tilde{c}_1 =: \tilde{c}_{1a} \quad (56)$$

where, similarly, the subscripts an stand for a, antisymmetrical (OP), and n, quantization order (nth quantization), and Z_{an} is the nth solution of the equation in (55) regarding the asymmetrical solutions. Similarly, by using the results in (13), we find the MAF quantization formula regarding the antisymmetrical solution:

$$\int_0^{y_2} k\left[\tilde{\lambda}_{M,an}, y\right] dy =: \left(\zeta_{an} + \frac{1}{2}\right)\pi \Rightarrow \zeta_{an} = \left(\frac{4Z_{an}^{3/2}}{3\pi} - \frac{1}{2}\right), n = 1, 2, 3, \ldots \quad (57)$$

where ζ_{an} is the universal MAF constants regarding the antisymmetrical solution whose values are given in **Table 2** along with the JWKB solutions (which are already exact) for some n values in comparison. Note that we used n_a to represent the antisymmetrical (OP) MAF indices in **Table 2**.

4.3 MAF solution of eigenfunctions

By using a tentative boundary condition with $q = 1$ for the EP solutions, we have found the result in (47), and we said that we would extend it by considering the parity matching for $q = \pm q$. Consequently,

$$\tilde{\psi}^{(\text{par.m.})}_{(M,E.P.)}(\lambda_{n_s}, y \geq 0) = (-1)^{\left(\frac{n_s}{2}\right)} \times \frac{\tilde{c}_{1s}}{\sqrt{\partial_y \xi(\lambda_n, y)}} Ai[\xi(\lambda_n, y)], n_s = 0, 2, 4, \ldots (even) \quad (58)$$

or

$$\tilde{\psi}^{(\text{par.m.})}_{(M,E.P.)}(\lambda_{n_s}, y \geq 0) = (-1)^{\left(\frac{n_s}{2}\right)} \times \sqrt{\frac{\xi_0'}{\beta \partial_y Ai[\xi(\lambda_{n_s}, y)]}\frac{Ai[\xi(\lambda_{n_s}, y)]}{Ai(\xi_0)}}, n_s = 0, 2, 4, \ldots (even) \quad (59)$$

Similarly, for the antisymmetric parity wave functions, we have.

$$\tilde{\psi}^{(\text{par.m.})}_{(M,O.P.)}(\lambda_{n_a}, y \geq 0) = (-1)^{\left(\frac{n_a-1}{2}\right)} \times \frac{\tilde{c}_{1a}}{\sqrt{\partial_y \xi(\lambda_n, y)}} Ai[\xi(\lambda_n, y)], n_s = 1, 3, 5, \ldots (odd) \quad (60)$$

where constant coefficients $\tilde{c}_{1s}$ and $\tilde{c}_{1a}$ represent the related symmetric and antisymmetric coefficients, respectively.

5. MAF solution of the QHO

Again, we use the schematic sketch given in **Figure 2** for the QHO under study.

5.1 MAF eigenenergies of the QHO

Since we have tactically used (53)–(57) to resemble the MAF quantization formula to the JWKB quantization formula given in (12), by using the result of calculation of the same integral in (13), we have the following results regarding the MAF eigenenergies of the QHO:

$$E.P.:\tilde{\lambda}_{M,sn}(E_{sn})=\sqrt{2\zeta_{sn}+1}(\text{or})\tilde{E}_{M,sn}=:E_{MAF,sn}=\left(\zeta_{sn}+\frac{1}{2}\right)\pi;n=1,2,3,\ldots \tag{61}$$

$$O.P.:\tilde{\lambda}_{M,an}(E_{an})=\sqrt{2\zeta_{an}+1}(\text{or})\tilde{E}_{M,an}=:E_{MAF,an}=\left(\zeta_{sn}+\frac{1}{2}\right)\pi;n=1,2,3,.. \tag{62}$$

MAF eigenenergies are given in **Table 2** along with the JWKB solutions (which are already exact) for some n values in comparison. (Note again that we used n_s and n_a to represent the symmetrical (EP) and antisymmetrical MAF indices in **Table 2,** respectively).

5.2 MAF eigenfunctions of the QHO

For the regions IIb and III, we have the following definitions:

$$f(\lambda_n,y)=k^2(\lambda_n,y)=\begin{cases}k^2(\tilde{\lambda}_M\leftarrow\tilde{\lambda}_n,y)=\tilde{\lambda}_n^2-y^2;0\le y\le\tilde{\lambda}_n\\ \kappa^2(\tilde{\lambda}_M\leftarrow\tilde{\lambda}_n,y)=y^2-\tilde{\lambda}_n^2;\tilde{\lambda}_n\le y<\infty\end{cases} \tag{63}$$

Calculation of ξ in (37) for the first quadrant gives.

$$\xi(\lambda_n,y)=\begin{cases}\xi_{IIb}:-\left[\frac{3}{2}\int_0^y k(\tilde{\lambda}_M,y)dy\right]^{2/3}=-\left[\frac{3}{2}\int_y^{y_{t2}}k(\tilde{\lambda}_M,y)dy\right]^{2/3},for\ 0\le y\le y_{t2}\\ \xi_{III}:\left[\frac{3}{2}\int_{y_{t2}}^y\kappa(\tilde{\lambda}_M,y)dy\right]^{2/3},for\ y_{t2}\le y\end{cases}$$

$$=\begin{cases}\xi_{IIb}:-\dfrac{3^{2/3}}{4}\left\{-2yk(\tilde{\lambda}_M,y)+\tilde{\lambda}_M^2\left[\pi-2\arctan\left(\dfrac{y}{k(\tilde{\lambda}_M,y)}\right)\right]\right\}^{2/3},for\ 0\le y\le\tilde{\lambda}_M\\ \xi_{III}:\dfrac{3^{2/3}\left\{y\kappa(\tilde{\lambda}_M,y)+\tilde{\lambda}_M^2\ln\left[\dfrac{\tilde{\lambda}_M}{y\kappa(\tilde{\lambda}_M,y)}\right]\right\}^{2/3}}{2^{4/3}},for\ \tilde{\lambda}_M\le y<\infty\end{cases} \tag{64}$$

where $\tilde{\lambda}_M$ (and also $\tilde{E}_M$ below) are MAF eigenenergies which become (61) for the symmetric (EP) and (62) for the antisymmetric (OP) case accordingly. Calculation of the constant coefficient in (58) or (59) for the symmetric boundary values given in (46) with q = 1 gives.

$$E.P.:\tilde{c}_{1_s}=\frac{\sqrt{2}\tilde{E}_{s_n}^{1/6}}{\sqrt{\beta}(3\pi)^{1/6}Ai\left[-\frac{1}{4}\tilde{E}_{s_n}^{4/3}(3\pi)^{2/3}\right]} \tag{65}$$

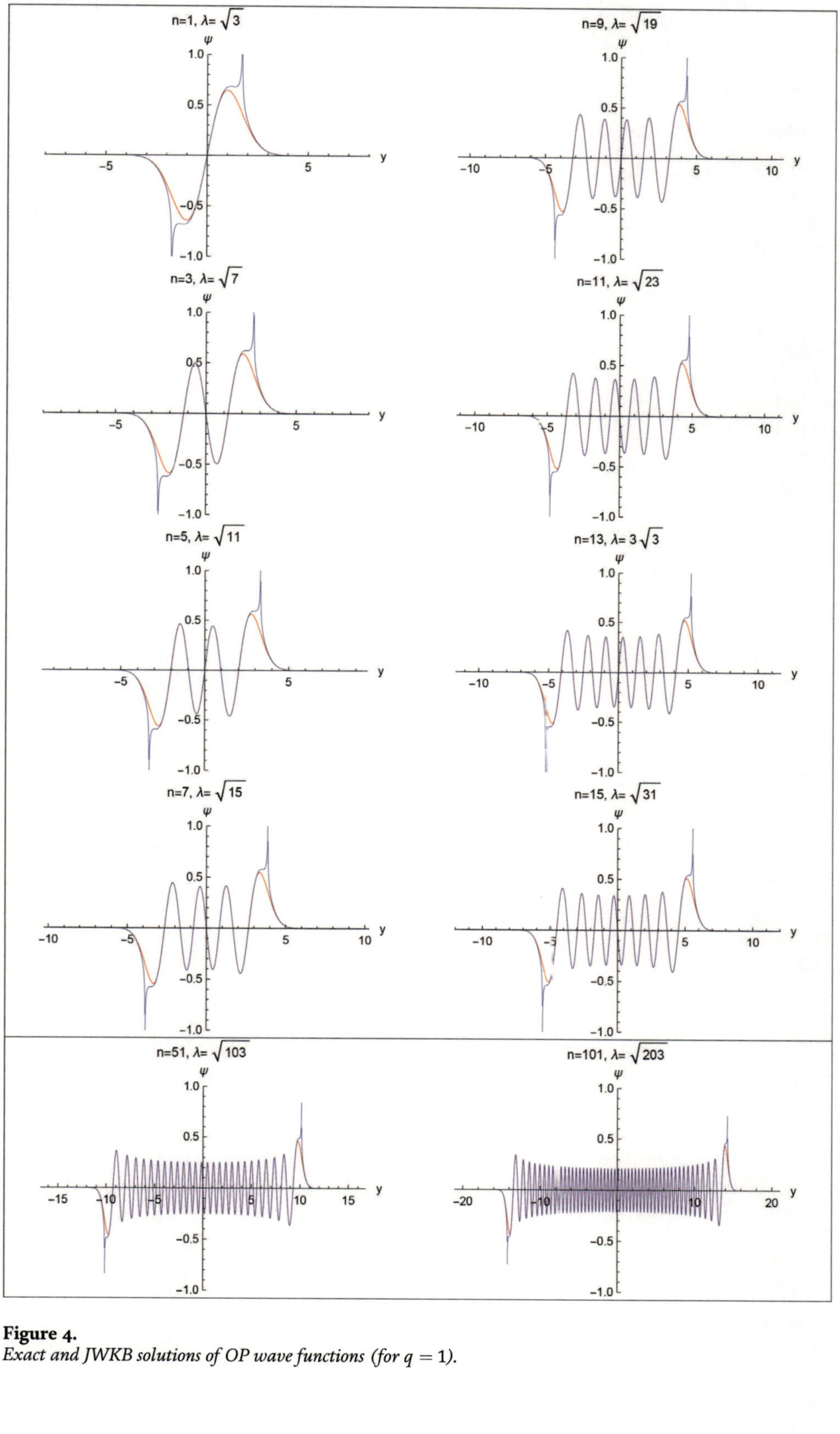

Figure 4.
Exact and JWKB solutions of OP wave functions (for $q = 1$).

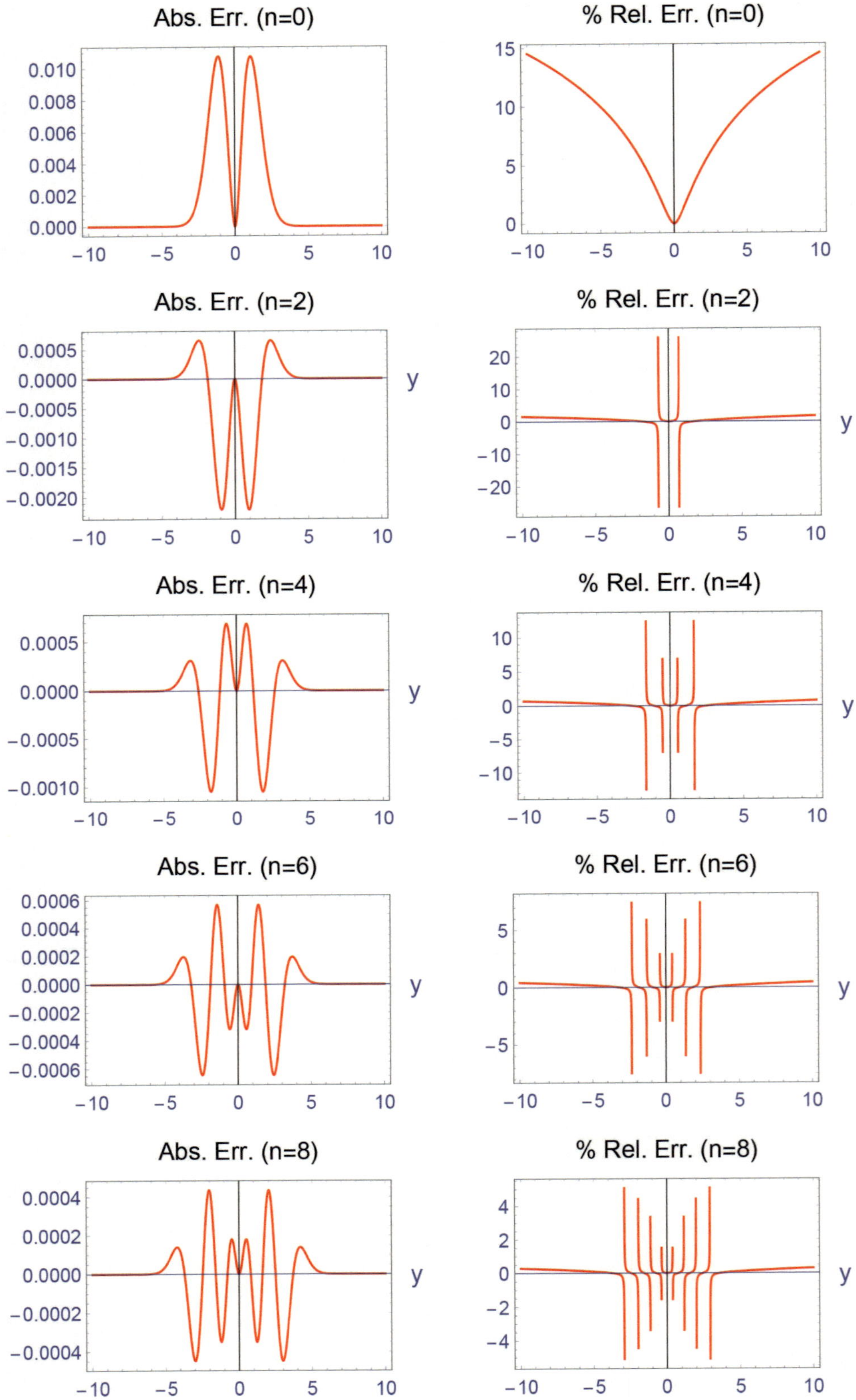

Figure 5.
Relative and absolute error of EP MAF solutions.

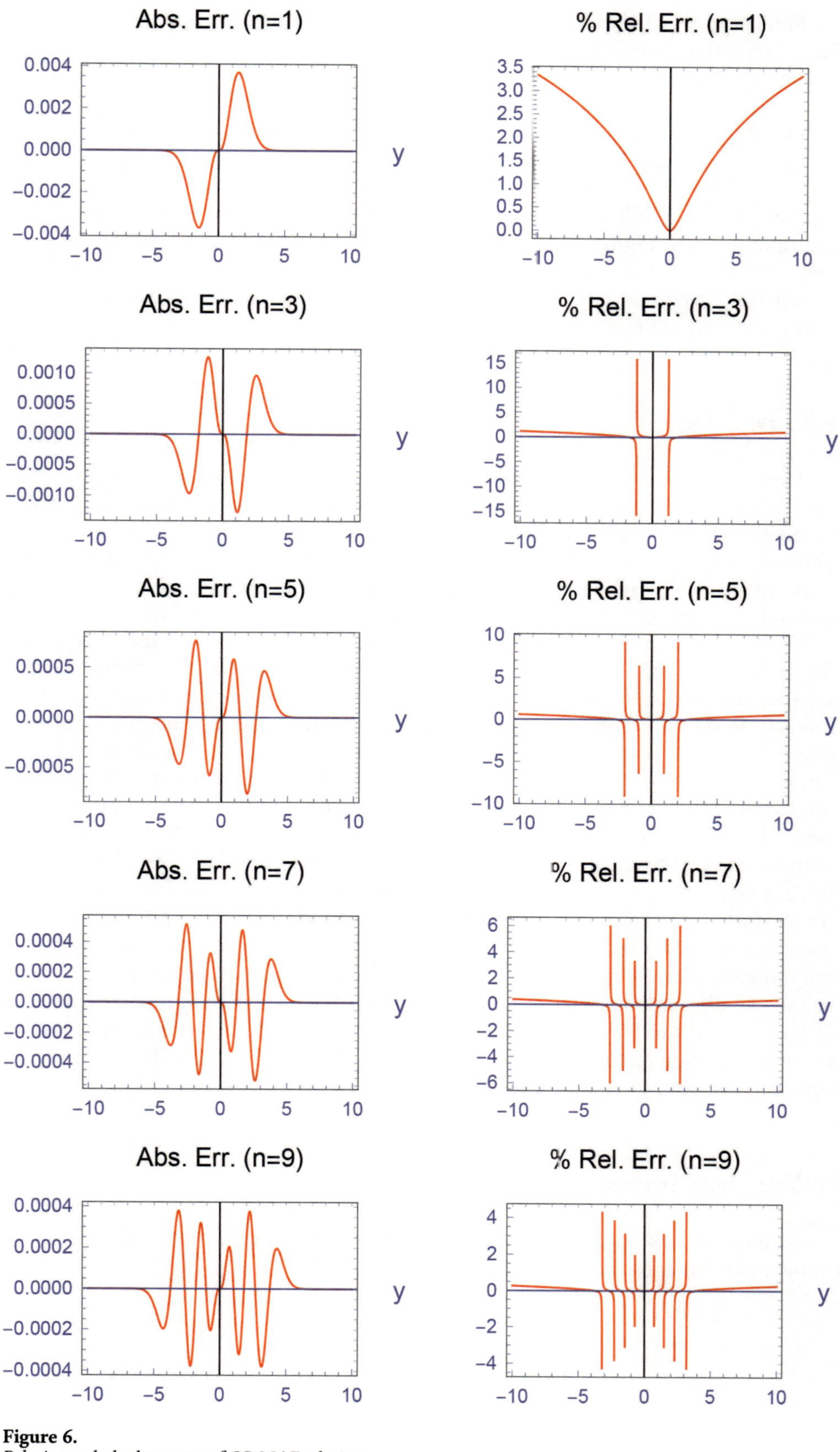

Figure 6.
Relative and absolute error of OP MAF solutions.

Similarly, calculation of the constant coefficient in (60) for the antisymmetric boundary values given in (54) with $q = 1$ gives.

$$O.P.:\tilde{c}_{1a} = \frac{\tilde{E}_{a_n}^{7/6}(3\pi)^{1/6}}{\sqrt{2\beta^3}\left\{\tilde{E}_{a_n}^{4/3}(3\pi)^{2/3}Ai'\left[-\frac{1}{4}\tilde{E}_{a_n}^{4/3}(3\pi)^{2/3}\right] - Ai\left[-\frac{1}{4}\tilde{E}_{a_n}^{4/3}(3\pi)^{2/3}\right]\right\}} \tag{66}$$

Since the MAF solutions of both EP and OP solutions are very close to the exact solutions given in **Figures 3** and **4**, their absolute and relative error graphs with respect to the exact solution are given in **Figures 5** and **6**. We can also see that there are no discontinuities at the CTPs in the MAF solutions when compared with the JWKB solutions given in **Figures 3** and **4**.

6. Conclusion

Here we studied the fundamental outcomes of the two conventional semiclassical approximation methods, namely, JWKB and MAF methods pedagogically, and obtained the solutions of the QHO by these semiclassical methods by using the parity conditions of the expected solutions by using the dimensionless form of the QHO system. We applied the asymptotic matching and parity matching procedure to obtain the correct form of semiclassical solutions. As expected, JWKB solutions diverge at and around the CTPs, whereas MAF solutions do not. As also expected (since being typical), JWKB eigenenergies are exact, whereas MAF eigenenergies are unfortunately not but very accurate as expected from an approximation method. In the MAF method, function p in (40) or in (43) is assumed zero. Indeed, it is very close to zero to give approximate results, and function P in (40) or in (43) can be used as an approximation criterion for the MAF method [3, 18]. However, improved MAF methods (IMAF) or perturbation corrections concerning the non-zero P function seem straightforward to improve the accuracy of the MAF solutions as in [3, 20, 22]. Normally, for an even potential function in the TISE, EP and OP initial values are as given in (17), but due to the conversion factor β in (11) or (16), for the QHO in the dimensionless form (in ψ), we have (22) and (29). In our notation, we have used the notation, $\varphi \Leftrightarrow \psi$, where real physical system is in φ and the dimensionless form is in ψ. Since the standard formulation is given according to the real physical systems, JWKB and MAF formulas in the literature such as in [1–7, 19–23] surely correspond to the initial values $\beta \rightarrow 1$ in our dimensionless form formulation in ψ. Consequently, we hereby present a full JWKB and MAF solution concerning the quantized conversion factor β in (11).

Acknowledgements

Author acknowledges special thanks to the IntechOpen for the financial support in the publishment of this chapter.

DOI: http://dx.doi.org/10.5772/intechopen.85147

Author details

Coşkun Deniz
Department of Electrical and Electronics Engineering, Faculty of Engineering, Adnan Menderes University, Turkey

*Address all correspondence to: cdeniz@adu.edu.tr

References

[1] Liboff R. Introductory Quantum Mechanics, 4/E. Cornell University: Addison Wesley; 2001

[2] Griffiths DJ. 20 Introduction to Quantum Mechanics, 2nd ed. Upper Saddle River, NJ: Pearson; 2005. ISBN: 0-13-191175-9

[3] Ghatak AK, Gallawa RL, Goyal IC. Modified Airy Functions and WKB Solutions to the Wave Equation. Washington: NIST; 1991

[4] Deniz C. Semiclassical anomalies of the quantum mechanical systems and their modifications for the asymptotic matching. Annals of Physics. 2011; **326**(8):1816-1838

[5] Landau LD, Lifshitz EM. Quantum mechanics, non-relativistic theory. In: Course of Theoretical Physics. 2nd ed. Vol. 3. NY: Pergamon; 1965

[6] Hruska M, Keung WY, Sukhatme U. Accuracy of semiclassical methods for shape-invariant potentials. Physical Review A. 1997;**55**(5):3345-3350

[7] Bender CM, Orszag SA. Advanced Mathematical Methods for Scientists and Engineers Asymptotic Methods and Perturbation Theory. NY: Springer–Verlag; 1999

[8] Aydin MC, Uncu H, Deniz C. A parabolic model for dimple potentials. Physica Scripta. 2013;**88**(3):035006. Available from: http://iopscience.iop.org/article/10.1088/0031-8949/88/03/035006/meta

[9] Gómez-Vergel D, Villaseñor E. The time-dependent quantum harmonic oscillator revisited: Applications to quantum field theory. Annals of Physics. 2009;**324**:1360-1385

[10] Tong T. Quantum Field Theory, University of Cambridge Part III. Mathematical Tripos. 2007. Available from: http://www.damtp.cam.ac.uk/user/tong/qft.html

[11] Gin-Ge Chen B, Derbes D, Griffiths D, Hill B, Sohn R, Ting YS. Lectures of Sidney Coleman on Quantum Field Theory. 1st ed. NJ: World Scientific; 2018

[12] Shankar R. Principles of Quantum Mechanics. Chapter 10, Exercises 10.2.2–10.2.3. 2nd ed. NY: Springer, Plenum Press; 1994. pp. 259-260. ISBN 0-306-44790-8

[13] Shankar R. Principles of Quantum Mechanics. Chapter 12. 2nd ed. NY: Springer, Plenum Press; 1994. pp. 351-352. ISBN 0-306-44790-8

[14] Deniz C. On the exact and JWKB solution of 1D quantum harmonic oscillator by mathematica. Journal of Physics: Conference Series. 2016; **707**(012033):1-11. Available from: h ttp://iopscience.iop.org/article/10.1088/1742-6596/707/1/012033/pdf

[15] Lindblad G. Brownian motion of a quantum harmonic oscillator. Reports on Mathematical Physics. 1976;**10**(3): 393-406

[16] Ford GW, Lewis JT, O'Connel RF. Quantum Langevin equation. Physical Review A. 1988;**37**(11):4419-4428

[17] Stenholm S. Quantum theory of linear friction. Brazillian Journal of Physics. 1997;**27**(2):214-237

[18] Deniz C, Gerceklilu M. An analysis of the exactly solvable linear differential equations by the modified airy function (MAF) method. Indian Journal of Physics. 2011;**85**(2):339-357. DOI: 10.1007/s12648-011-0010-1. Available from: http://link.springer.com/article/10.1007/s12648-011-0010-1

[19] Goyal IC, Gallawa RL, Ghatak AK. Improved variational analysis of inhomogenous optical waveguides using airy functions. Journal of Lightwave Technology. 1993;**11**(10):1575-1578

[20] Goyal IC, Rajeev J, Ghatak AJ. Planar optical waveguides with arbitrary index profile: An accurate method of analysis. Journal of Lightwave Technology. 1997;**15**(11):2179-2182

[21] Ghatak AK, Goyal IC, Jindal R, Varshni YP. MAF solution for bounded potential problems. Canadian Journal of Physics. 1998;**76**:351-359

[22] Goyal IC. Modified airy function solutions to optical waveguide problems. In: Proceedings of the 2002 4th International Conference on Transparent Optical Networks; IEE. 1. 2002. pp. 155-1605. Also available by print ISBN: 0-7803-7375-8/02

[23] Deniz C. On the MAF solution of the uniformly lengthening pendulum via change of independent variable in the Bessels equation. Results in Physics. 2017;**733**:333-343. Available from: https://www.sciencedirect.com/science/article/pii/S2211379717314092

[24] Boccara N. Essentials of Mathematica with Applications to Mathematics and Physics. NY: Springer; 2007. pp. 475-480

[25] Langer RE. On the asymptotic solutions of ordinary differential equations, with an application to the bessel functions of large order. The Transactions of the American Mathematical Society. 1931;**33**:23-64

Chapter 3

Oscillation Criteria of Two-Dimensional Time-Scale Systems

Ozkan Ozturk

Abstract

Oscillation and nonoscillation theories have recently gotten too much attention and play a very important role in the theory of time-scale systems to have enough information about the long-time behavior of nonlinear systems. Some applications of such systems in discrete and continuous cases arise in control and stability theories for the unmanned aerial and ground vehicles (UAVs and UGVs). We deal with a two-dimensional nonlinear system to investigate the oscillatory behaviors of solutions. This helps us understand the limiting behavior of such solutions and contributes several theoretical results to the literature.

Keywords: oscillation, nonoscillation, two-dimensional systems, time scale, nonlinear system, fixed point theorems

1. Introduction

This chapter analyses the oscillatory behavior of solutions of two-dimensional (2D) nonlinear time-scale systems of first-order dynamic equations. We also investigate the existence and asymptotic properties of such solutions. The tools that we use are the most well-known fixed point theorems to consider the sign of the component functions of solutions of our system. A *time scale*, denoted by $\mathbb{T}$, is an arbitrary nonempty closed subset of the real numbers $\mathbb{R}$, which is introduced by a German mathematician, Stefan Hilger, in his PhD thesis in 1988 [1]. His primary purpose was to unify continuous and discrete analysis and extend the results to one comprehensive theory. For example, the results hold for differential equations when $\mathbb{T} = \mathbb{R}$, while the results hold for difference equations when $\mathbb{T} = \mathbb{Z}$. Therefore, there might happen to be two different proofs and maybe similar in most cases. In other words, our essential desire is to combine continuous and discrete cases in one comprehensive theory and remove the obscurity from both. For more details in the theory of differential and difference equations, we refer the books [2–4] to interested readers. As for the time-scale theory, we assume most of the readers are not familiar with the time-scale calculus, and thus we give a concise introduction to the theory of time scales from the books [5, 6] written by Bohner and Peterson in 2001 and 2003, respectively.

Two-dimensional dynamical systems have recently gotten too much attention because of their potential in applications in engineering, biology, and physics (see, e.g., [7–11]). For example, Bartolini and Pvdvnowski [12] consider a nonlinear system and propose a new method for the asymptotic linearization by means of continuous control law. Also Bartolini et al. [13, 14] consider an uncertain second-order nonlinear system and propose a new approximate linearization and sliding mode to control such systems. In addition to the nonoscillation for two-dimensional systems of first-order equations, periodic and subharmonic solutions are also investigated in [15–17], and significant contributions have been made. Another type of two-dimensional systems of dynamic equations is the Emden-Fowler type equation, named after E. Fowler after he did the mathematical foundation of a second-order differential equation in a series of four papers during 1914–1931 (see [18–21]). This system has several fascinating applications such as in gas dynamics and fluid mechanics, astrophysics, nuclear physics, relativistic mechanics, and chemically reacting systems (see [9, 22–24]).

This chapter is organized as follows: In Section 2, we give the calculus of the time-scale theory for those who are not familiar with the time scale (see [5]). In Section 3, referred to [25, 26], we show the existence and asymptotic behaviors of nonoscillatory solutions of a two-dimensional homogeneous dynamical system on time scales by using improper integrals and some inequalities. We also give enough examples for readers to see our results work nicely. Section 4, referred to [27], provides us oscillation criteria for two-dimensional nonhomogeneous time-scale systems by using famous inequalities and rules such as comparison theorem and chain rules on time scales. Finally, we give a conclusion and provide some exercises to the readers to have them comprehend the main results in the last two sections.

2. Preliminaries

The examples of the time scales are not restricted with the set of real numbers $\mathbb{R}$ and the set of integers $\mathbb{Z}$. There are several other time scales which are used in many application areas such as $q^{\mathbb{N}_0} = \{1, q, q^2, \cdots,\}, q > 1$ (called q-difference equations [28]), $\mathbb{T} = h\mathbb{Z}, h > 0$, $\mathbb{T} = \mathbb{N}_0^2 = \{n^2 : n \in \mathbb{N}_0\}$, etc. On the other hand, the set of rational numbers $\mathbb{Q}$, the set of irrational numbers $\mathbb{R}\backslash\mathbb{Q}$, and the open interval (a, b) are not time scales since they are not closed subsets of $\mathbb{R}$. For the following definitions and theorems in this section, we refer [5], (Chapter 1), and [29] to the readers.

Definition 2.1 Let $\mathbb{T}$ be a time scale. Then, the *forward jump operator* $\sigma : \mathbb{T} \to \mathbb{T}$ is defined by

$$\sigma(t) := \inf\{s \in \mathbb{T} : s > t\} \quad \text{for all} \quad t \in \mathbb{T}$$

while the *backward jump operator* $\rho : \mathbb{T} \to \mathbb{T}$ is given by

$$\rho(t) := \sup\{s \in \mathbb{T} : s < t\} \quad \text{for all} \quad t \in \mathbb{T}.$$

Finally, the *graininess function* $\mu : \mathbb{T} \to [0, \infty)$ is defined by $\mu(t) := \sigma(t) - t \quad \text{for all} \quad t \in \mathbb{T}$.

For a better explanation, the operator σ is the first next point, while the operator ρ is the first back point on a time scale. And μ is the length between the next point and the current point. So it is always nonnegative. **Table 1** shows some examples of the forward/backward jump operators and the graininess function for most known time scales.

If $t < \sup\mathbb{T}$ and $\sigma(t) = t$, then t is said to be *right-dense*, and if $t > \inf\mathbb{T}$ and $\rho(t) = t$, we say t is *left-dense*. Also, if t is right- and left-dense at the same time, then t is said to be *dense*. In addition to left and right-dense points, it is said to be *right-scattered* when $\sigma(t) > t$, and t is called *left-scattered* when $\rho(t) < t$. Also, if t is right-and left-scattered at the same time, then t is called *isolated*. **Figure 1** shows the classification of points on time scales, clarifying the operators σ, ρ and μ (see [5]).

Next, we introduce the definition of derivative on any time scale. Note that if $\sup\mathbb{T} < \infty$, then $\mathbb{T}^\kappa = \mathbb{T} \backslash (\rho(\sup\mathbb{T}), \sup\mathbb{T}]$, and $\mathbb{T}^\kappa = \mathbb{T}$ if $\sup\mathbb{T} = \infty$. Suppose that $f : \mathbb{T} \to \mathbb{R}$ is a function. Then $f^\sigma : \mathbb{T} \to \mathbb{R}$ is defined by $f^\sigma(t) = f(\sigma(t)) \quad for \ all \quad t \in \mathbb{T}$.

Definition 2.2 If there does exist a $\delta > 0$ such that

$$|g(\sigma(t)) - g(s) - g^\Delta(t)(\sigma(t) - s)| \leq \varepsilon|\sigma(t) - s| \quad for \ all \quad s \in (t - \delta, t + \delta) \cap \mathbb{T},$$

for any ε, then g is called *delta differentiable* on $\mathbb{T}^\kappa$ and g^Δ is said to be *delta derivative* of g. Sometimes, delta derivative is referred as Hilger derivative in the literature (see [5]).

Theorem 2.3 *Suppose that $f, g : \mathbb{T} \to \mathbb{R}$ is a function with $t \in \mathbb{T}^\kappa$. Then.*

i. g is said to be continuous at t if g is differentiable at t.

ii. g is differentiable at t and

$$g^\Delta(t) = \frac{g(\sigma(t)) - g(t)}{\mu(t)},$$

provided g is continuous at t and t is right-scattered.

iii. Let t be right-dense, then g is differentiable at t if and only if

$$g^\Delta(t) = \lim_{s \to t} \frac{g(t) - g(s)}{t - s}$$

is equal to a finite number.

$\mathbb{T}$	$\sigma(t)$	$\rho(t)$	$\mu(t)$
$\mathbb{R}$	t	t	0
$h\mathbb{Z}$	$t + h$	$t - h$	h
$\mathbb{N}_0^2$	$(\sqrt{t} + 1)^2$		$1 + 2\sqrt{t}$
$q^{\mathbb{N}_0}$	tq	$\frac{t}{q}$	$(q - 1)t$

Table 1.
Examples of most known time scales.

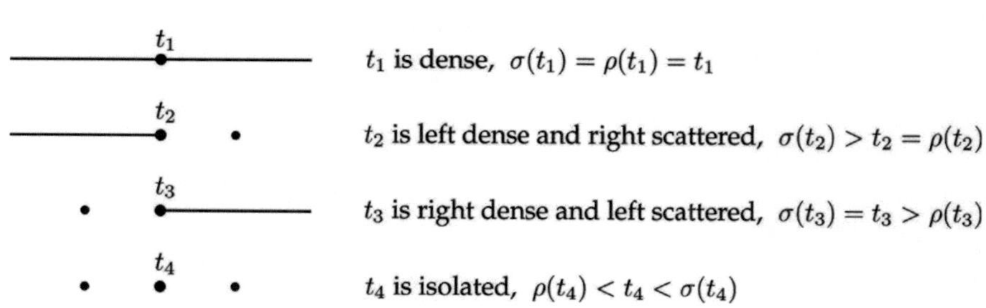

Figure 1.
Classification of points.

iv. If $g(t)g(\sigma(t)) \neq 0$, then $\frac{f}{g}$ is differentiable at t with

$$\left(\frac{f}{g}\right)^{\Delta}(t) = \frac{f^{\Delta}(t)g(t) - f(t)g^{\Delta}(t)}{g(t)g(\sigma(t))}.$$

If $\mathbb{T} = \mathbb{R}$, then f^{Δ} turns out to be the usual derivative f' on continuous case, while f^{Δ} is reduced to forward difference operator Δf, defined by $\Delta f(t) = f(t+1) - f(t)$ if $\mathbb{T} = \mathbb{Z}$. The following example is a good example of time scale applications in electrical engineering (see [5], Example 1.39–1.40).

Example 2.4 Consider a simple electric circuit, shown in **Figure 2** with resistor R, inductor L, capacitor C and the current I.

Suppose, we discharge the capacitor periodically every time unit and assume that the discharging small $\delta > 0$ time units. Then we can model it as

$$\mathbb{P}_{1-\delta,\delta} = \bigcup_{k \in \mathbb{N}_0} [k, k+1-\delta]$$

by using the time scale. Suppose that $Q(t)$ is the total charge on the capacitor at time t and $I(t)$ is the current with respect to time t. Then the total charge Q can be defined by

$$Q^{\Delta}(t) = \begin{cases} bQ(t) & \text{if } \; t \in \bigcup_{k \in \mathbb{N}} \{k - \delta\} \\ I & \text{otherwise} \end{cases}$$

and

$$I^{\Delta}(t) = \begin{cases} 0 & \text{if } \; t \in \bigcup_{k \in \mathbb{N}} \{k - \delta\} \\ -\frac{1}{LC}Q(T) - \frac{R}{L}I(t) & \text{otherwise,} \end{cases}$$

where $-1 < b\delta < 0$.

Finally, we introduce the integrals on time scales, but before that, we must give the following definition to define delta integrable functions (see [5]).

Definition 2.5 $g : \mathbb{T} \to \mathbb{R}$ is said to be *right-dense continuous* (rd-continuous) if its left-sided limits exist at left-dense points in $\mathbb{T}$ and it is continuous at right-dense points in $\mathbb{T}$. We denote rd-continuous functions by $C_{rd}(\mathbb{T}, \mathbb{R})$. The set of functions g that are differentiable and whose derivative is *rd-continuous* is denoted by $C^1_{rd}(\mathbb{T}, \mathbb{R})$. Finally, we denote continuous functions by C throughout this chapter.

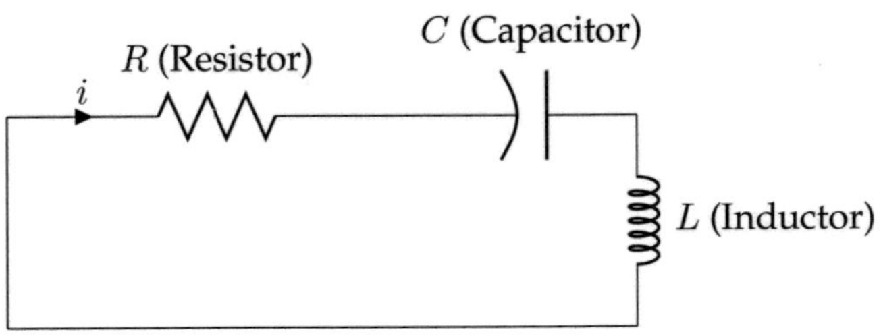

Figure 2.
Electric circuit.

Theorem 2.6 ([5], Theorem 1.60) *For $\sigma : \mathbb{T} \to \mathbb{T}$ and $f : \mathbb{T} \to \mathbb{R}$, we have the following:*

i. *The jump operator σ is rd-continuous.*

ii. *If f is continuous, then it is rd-continuous.*

The Cauchy integral is defined by

$$\int_a^b f(t)\Delta t = F(b) - F(a) \quad \text{for all} \quad a, b \in \mathbb{T}.$$

The following theorem presents the existence of antiderivatives.

Theorem 2.7 *Every rd-continuous function has an antiderivative. Moreover, F given by*

$$F(t) = \int_{t_0}^t f(s)\Delta s \quad \textit{for} \quad t \in \mathbb{T}$$

is an antiderivative of f.

Similar to the continuous analysis, we have integral properties and some of them are presented as follows ([5] or [29]):

Theorem 2.8 *Suppose that h_1 and h_2 are rd-continuous functions, $c, d, e \in \mathbb{T}$ and $\beta \in \mathbb{R}$.*

i. h_1 is nondecreasing if $h_1^\Delta \geq 0$.

ii. If $h_1(t) \geq 0$ for all $c \leq t \leq d$, then $\int_c^d h_1(t)\Delta t \geq 0$.

iii. $\int_c^d [(\beta h_1(t)) + (\beta h_2(t))] = \beta \int_c^d h_1(t)\Delta t + \beta \int_a^b h_2(t)\Delta t$.

iv. $\int_c^e h_1(t)\Delta t = \int_c^d h_1(t)\Delta t + \int_d^e h_1(t)\Delta t$.

v. $\int_c^d h_1(t)h_2^\Delta(t)\Delta t = (h_1h_2)(d) - (h_1h_2)(c) - \int_c^d h_1^\Delta(t)h_2(\sigma(t))\Delta t$

vi. $\int_a^a f(t)\Delta t = 0$.

Table 2 shows how the derivative and integral are defined for some time scales for $a, b \in \mathbb{T}$.

$\mathbb{T}$	$f^\Delta(t)$	$\int_a^b f(t)\Delta t$
$\mathbb{R}$	$f'(t)$	$\int_a^b f(t)dt$
$\mathbb{Z}$	$\Delta f(t)$	$\sum_{t=a}^{b-1} f(t)$
$q^{\mathbb{N}_0}$	$\Delta_q f(t)$	$\sum_{t \in [a,b)_{q^{\mathbb{N}_0}}} f(t)\mu(t)$

Table 2.
Derivative and integrals for most common time scales.

We finish the section by Schauder's fixed point theorem, proved by Juliusz Schauder in 1930, and Knaster fixed point theorem, proved by Knaster in 1928 (see [30], Theorem 2.A and [31], respectively).

Theorem 2.9 Schauder's fixed point theorem. *Suppose that S is a nonempty, bounded, closed, and convex subset of a Banach space Y and that* $F : S \to S$ *is a compact operator. Then, we conclude that F has a fixed point such that* $y = Fy$.

Theorem 2.10 The Knaster fixed point theorem. *Suppose that* $(S, \leq)$ *is a complete lattice and that* $F : S \to S$ *is order preserving, then F has a fixed point such that* $y = Fy$. *In fact, we say that the set of fixed points of F is a complete lattice.*

Finally, we note that throughout this paper, we assume that $\mathbb{T}$ is unbounded above and whenever we write $t \geq t_1$, we mean $t \in [t_1, \infty)_{\mathbb{T}} := [t_1, \infty) \cap \mathbb{T}$.

3. Nonoscillation on a two-dimensional time-scale systems

This section focuses on the nonoscillatory solutions of a two-dimensional dynamical system on time scales. To do this, we consider the system

$$\begin{cases} x^{\Delta}(t) = p(t)f(y(t)) \\ y^{\Delta}(t) = r(t)g(x(t)), \end{cases} \tag{1}$$

where $p, r \in C_{rd}([t_0, \infty)_{\mathbb{T}}, \mathbb{R}^+)$ and f and g are nondecreasing functions such that $uf(u) > 0$ and $ug(u) > 0$ for $u \neq 0$.

By a solution of (1), we mean a collection of functions, where $x, y \in C^1_{rd}([t_0, \infty)_T; R)$, $T \geq t_0$ and (x, y) satisfies system (1) for all large $t \geq T$.

Note that system (1) is reduced to the system of differential equations when the time scale is the set of real numbers $\mathbb{R}$, i.e., $f^{\Delta} = f'$ (see [32]). And when $\mathbb{T} = \mathbb{Z}$, system (1) turns out to be a system of difference equations, i.e., $f^{\Delta} = \Delta f$ (see [33]). Other versions of system (1), the case $\mathbb{T} = \mathbb{Z}$, are investigated by Li et al. [34], Cheng et al. [35], and Marini et al. [36]. More details about the continuous and discrete versions of system (1) are given in the conclusion section.

Definition 3.1 A solution (x, y) of system (1) is said to be proper if

$$\sup\{|x(s)|, |y(s)|, |z(s)| : s \in [t, \infty)_{\mathbb{T}}\} > 0$$

holds for $t \geq t_0$.

Definition 3.2 A proper solution (x, y) of (1) is said to be nonoscillatory if the component functions x and y are both nonoscillatory, i.e., either eventually positive or eventually negative. Otherwise it is said to be oscillatory.

Suppose that N is the set of all nonoscillatory solutions of system (1). It can easily be shown that any nonoscillatory solution (x, y) of system (1) belongs to one of the following classes:

$$N^+ := \{(x, y) \in N : xy > 0 \text{ eventually}\}$$

$$N^- := \{(x, y) \in N : xy < 0 \text{ eventually}\}.$$

Let (x, y) be a solution of system (1). Then one can show that the component functions x and y are themselves nonoscillatory (see, e.g., [37]). Throughout this section, we assume that the first component function x of the nonoscillatory solution (x, y) is eventually positive. The results can be obtained similarly for the case $x < 0$ eventually.

We obtain the existence criteria for nonoscillatory solutions of system (1) in N^+ and N^- by using the fixed point theorems and the following improper integrals:

$$I_1 = \int_{t_0}^{\infty} p(t) f\left(k_1 \int_{t_0}^{t} r(s)\Delta s\right)\Delta t, \qquad I_2 = \int_{t_0}^{\infty} r(t) g\left(k_2 \int_{t_0}^{t} p(s)\Delta s\right)\Delta t,$$

$$I_3 = \int_{t_0}^{\infty} p(t) f\left(k_3 - k_4 \int_{t}^{\infty} r(s)\Delta s\right)\Delta t, \quad I_4 = \int_{t_0}^{\infty} r(t) g\left(k_5 \int_{t}^{\infty} p(s)\Delta s\right)\Delta t,$$

$$P(t_0, t) = \int_{t_0}^{t} p(s)\Delta s, \qquad R(t_0, t) = \int_{t_0}^{t} r(s)\Delta s,$$

where k_i, $i = 1 - 5$ are some constants.

3.1 Existence of nonoscillatory solutions of (1) in N^+

Suppose that (x, y) is a nonoscillatory solution of (1) such that $x > 0$. Then system (1) implies that $x^{\Delta} > 0$ and $y^{\Delta} > 0$ eventually. Therefore, as a result of this, we have that x converges to a positive finite number or $x \to \infty$ and similarly y tends to a positive finite number or $y \to \infty$. One can have very similar asymptotic behaviors when $x < 0$. Hence, as a result of this information, the following subclasses of N^+ are obtained:

$$N^+_{F,F} = \left\{(x,y) \in N^+ : \lim_{t\to\infty} |x(t)| = c, \quad \lim_{t\to\infty} |y(t)| = d\right\},$$

$$N^+_{F,\infty} = \left\{(x,y) \in N^+ : \lim_{t\to\infty} |x(t)| = c, \quad \lim_{t\to\infty} |y(t)| = \infty\right\},$$

$$N^+_{\infty,F} = \left\{(x,y) \in N^+ : \lim_{t\to\infty} |x(t)| = \infty, \quad \lim_{t\to\infty} |y(t)| = d\right\},$$

$$N^+_{\infty,\infty} = \left\{(x,y) \in N^+ : \lim_{t\to\infty} |x(t)| = \infty, \quad \lim_{t\to\infty} |y(t)| = \infty\right\}.$$

To focus on N^+, first consider the following four cases for $t_0 \in \mathbb{T}$:

1. $P(t_0, \infty) = \infty$ and $R(t_0, \infty) = \infty$

2. $P(t_0, \infty) = \infty$ and $R(t_0, \infty) < \infty$

3. $P(t_0, \infty) < \infty$ and $R(t_0, \infty) < \infty$

4. $P(t_0, \infty) < \infty$ and $R(t_0, \infty) = \infty$

Suppose $P(t_0, \infty) = \infty$ and $R(t_0, \infty) = \infty$ and that (x, y) is a nonoscillatory solution in N^+. Integrating the equations of system (1) from t_0 to t separately gives us

$$x(t) \geq x(t_0) + f(y(t_0)) \int_{t_0}^{t} p(s)\Delta s$$

and

$$y(t) \geq y(t_0) + g(x(t_0)) \int_{t_0}^{t} r(s)\Delta s, \quad t \geq t_0.$$

Thus, we get $x(t) \to \infty$ and $y(t) \to \infty$ as $t \to \infty$. In view of this information, the following theorem is given without any proof.

Theorem 3.3 *Let $P(t_0, \infty) = \infty$ and $R(t_0, \infty) = \infty$. Then any nonoscillatory solution of system (1) belongs to $N^+_{\infty,\infty}$.*

Next, we consider the other three cases to obtain the nonoscillation criteria for system (1).

3.1.1 The case $P(t_0, \infty) = \infty$ and $R(t_0, \infty) < \infty$

Suppose that (x, y) is a nonoscillatory solution of system (1) such that $x > 0$ and $y > 0$ eventually. Then by the integration of the first equation of system (1) from t_0 to t, we have that there exists $k > 0$

$$x(t) \geq x(t_0) + k \int_{t_0}^{t} p(s)\Delta s, \quad t_0 \in \mathbb{T}. \tag{2}$$

Then by taking the limit of (2) as $t \to \infty$, we have that x diverges. Therefore, we have the following lemma in the light of this information.

Lemma 3.4 *Any nonoscillatory solution in N^+ belongs to $N^+_{\infty,F}$, or $N^+_{\infty,\infty}$ for $0 < c, d < \infty$.*

It is not easy to give the sufficient conditions for the existence of nonoscillatory solutions in $N^+_{\infty,\infty}$. So, we only provide the existence of nonoscillatory solutions in $N^+_{\infty,F}$.

Theorem 3.5 *There exists a nonoscillatory solution in $N^+_{\infty,F}$ if and only if $I_2 < \infty$ for all $k_2 > 0$.*

Proof. Suppose that there exists a solution in $N^+_{\infty,F}$ such that $x(t) > 0, y(t) > 0$ for $t \geq t_0$, $x(t) \to \infty$ and $y(t) \to d$ as $t \to \infty$ for $d > 0$. Since y is eventually increasing, there exist $k_2 > 0$ and $t_1 \geq t_0$ such that $f(y(t)) \geq k_2$ for $t \geq t_1$. Integrating the first equation from t_1 to t, the monotonicity of f yields us

$$x(t) = x(t_1) + \int_{t_1}^{t} p(s)f(y(s))\Delta s \geq k_2 \int_{t_1}^{t} p(s)\Delta s, \quad t \geq t_1. \tag{3}$$

Integrating the second equation from t_1 to t, the monotonicity of g and (3) gives us

$$y(t) = y(t_1) + \int_{t_1}^{t} r(s)g(x(s))\Delta s \geq \int_{t_1}^{t} r(s)g\left(k_2 \int_{t_1}^{s} p(u)\Delta u\right)\Delta s, \quad t \geq t_1. \tag{4}$$

So as $t \to \infty$, we have that $I_2 < \infty$ holds.

Conversely, suppose that $I_2 < \infty$ for all $k_2 > 0$. Then, there exists a large $t_1 \geq t_0$ such that

$$\int_{t_1}^{\infty} r(t)g\left(k_2 \int_{t_1}^{t} p(s)\Delta s\right)\Delta t < \frac{c}{2}, \tag{5}$$

where $k_2 = f(c)$. Let Y be the set of all bounded and continuous real-valued functions $y(t)$ on $[t_1, \infty)_{\mathbb{T}}$ with the supremum norm $\sup_{t \geq t_1} |y(t)|$. Then Y is a Banach space (see [38]). Let us define a subset Ω of Y such that

$$\Omega := \left\{ y(t) \in Y : \frac{c}{2} \leq y(t) \leq c, \quad t \geq t_1 \right\}.$$

One can prove that Ω is bounded, closed, and also convex subset of Y. Suppose that $T : \Omega \to Y$ is an operator given by

$$(Ty)(t) = c - \int_t^\infty r(s) g\left(\int_{t_1}^s p(u) f(y(u)) \Delta u \right) \Delta s. \qquad (6)$$

The very first thing we do is to show that T is mapping into itself, i.e., $T : \Omega \to \Omega$.

$$\frac{c}{2} \leq c - \int_t^\infty r(s) g\left(\int_{t_1}^s p(u) f(c) \Delta u \right) \Delta s \leq (Ty)(t) \leq c$$

by using (5) for $y \in \Omega$. The second thing we show that T must be continuous on Ω. Hence, for $y \in \Omega$, suppose that y_n is a sequence in Ω so that $\|y_n - y\| \to 0$. Then

$$\begin{aligned} &|(Ty_n)(t) - (Ty)(t)| \\ &\leq \textstyle\int_t^\infty r(s) \left| g\left(\int_{t_1}^s p(u) f(y_n(u)) \Delta u \right) - g\left(\int_{t_1}^s p(u) f(y(u)) \Delta u \right) \right| \Delta s \end{aligned}.$$

Then by the Lebesgue dominated convergence theorem and by the continuity of f and g, we have that $\|Ty_n - Ty\| \to 0$ as $n \to \infty$, i.e., T, is continuous. Finally, we show that $T\Omega$ is relatively compact, i.e., equibounded and equicontinuous. Since

$$0 < (Ty)^\Delta(t) = r(t) g\left(\int_{t_1}^t p(u) f(y(u)) \Delta u \right) \leq r(t) g\left(k_2 \int_{t_1}^t p(u) \Delta u \right) < \infty,$$

we have that Ty is relatively compact by the Arzelá-Ascoli and mean value theorems. Therefore, Theorem 2.9 implies that there exists $\overline{y} \in \Omega$ such that $\overline{y} = T\overline{y}$. Then we have

$$\overline{y}^\Delta(t) = (T\overline{y})^\Delta(t) = r(t) g\left(\int_{t_1}^t p(u) f(\overline{y}(u)) \Delta u \right) \quad t \geq t_1. \qquad (7)$$

Setting $\overline{x}(t) = \int_{t_1}^t p(u) f(\overline{y}(u)) \Delta u$ gives us $x^\Delta(t) = p(t) f(\overline{y}(t))$. Hence, we have that $(\overline{x}, \overline{y})$ is a nonoscillatory solution of system (1) such that $\overline{x}(t) \to \infty$ and $\overline{y}(t) \to c$ as $t \to \infty$, i.e., $N^+_{\infty, F} \neq \emptyset$.

3.1.2 The case $P(t_0, \infty) < \infty$ and $R(t_0, \infty) < \infty$

In this subsection, we show that the existence of nonoscillatory solutions of (1) is only possible in $N^+_{F,F}$ and $N^+_{\infty,\infty}$ for $P(t_0, \infty) < \infty$ and $R(t_0, \infty) < \infty$, i.e., $N^+_{F,\infty} = N^+_{\infty,F} = \emptyset$.

Lemma 3.6 *Suppose $P(t_0, \infty) < \infty$ and $R(t_0, \infty) < \infty$ and that (x, y) is a nonoscillatory solution of system (1). Then $x(t)$ tends to a finite nonzero number c if and only if $y(t)$ tends to a finite nonzero number d as $t \to \infty$.*

Proof. We prove the theorem by assuming $x > 0$ without loss of generality. Therefore by the definition of N^+, y is also a positive component function of the solution (x,y). By taking the integral of the second equation of system (1) from t_0 to t and by the monotonicity of g and x, we have that there exists a positive constant k such that

$$y(t) \leq y(t_0) + k \int_{t_0}^{t} r(s)\Delta s,$$

where $k = g(c)$. Then we have that y is convergent because $P(t_0, \infty) < \infty$ as $t \to \infty$. The sufficiency can be shown similarly.

Theorem 3.7 $N^+_{F,F} \neq \emptyset$ *if and only if* $I_1 < \infty$ *for all* $k_1 > 0$.

Proof. The necessity part can be shown similar to Theorem 3.5. So for sufficiency, suppose $I_1 < \infty$ holds for all $k_1 > 0$. Then choose $t_1 \geq t_0$ such that

$$\int_{t_1}^{\infty} p(t)f\left(k_1 \int_{t_1}^{t} r(s)\Delta s\right)\Delta t < \frac{c}{2}, \tag{8}$$

where $k_1 = g(c)$ and $t \geq t_1$. Let X be the Banach space of all bounded real-valued and continuous functions on $[t_0, \infty)_{\mathbb{T}}$ with usual pointwise ordering $\leq$ and the norm $\sup_{t \geq t_1} |x(t)|$. Let Y be a subset of X such that

$$Y := \left\{x \in X : \frac{c}{2} \leq x(t) \leq c \quad t \geq t_1\right\}$$

and $F : \Omega \to X$ be an operator such that

$$(Fx)(t) = \frac{c}{2} + \int_{t_1}^{t} p(s)f\left(\int_{t_1}^{s} r(u)g(x(u))\Delta u\right)\Delta t, \quad t \geq t_1.$$

One can easily have that $\inf B \in Y$ and $\sup B \in Y$ for any subset B of Y, which implies that $(Y, \leq)$ is a complete lattice. First, let us show that $F : Y \to Y$ is an increasing mapping.

$$\frac{c}{2} \leq (Fx)(t) \leq \frac{c}{2} + \int_{t_1}^{t} p(s)f\left(g(c) \int_{t_1}^{s} r(u)\Delta u\right)\Delta t \leq c, \quad t \geq t_1,$$

that is $F : Y \to Y$. Note also that for $x_1 \leq x_2$, $x_1, x_2 \in Y$, we have $Fx_1 \leq Fx_2$, i.e., F, which is an increasing mapping. Then by Theorem 2.10, there exists a function $\overline{x} \in Y$ such that $\overline{x} = F\overline{x}$. By taking the derivative of $F\overline{x}$, we have

$$(F\overline{x})^{\Delta}(t) = p(t)f\left(\int_{t_1}^{t} r(u)g(\overline{x}(u))\Delta u\right), \quad t \geq t_1.$$

By letting

$$\overline{y}(t) = \int_{t_1}^{t} r(u)g(\overline{x}(u))\Delta u,$$

we have $\overline{y}^{\Delta}(t) = r(t)g(\overline{x}(t))$, and $(\overline{x}, \overline{y})$ is a nonoscillatory solution of system (1) such that $\overline{x}$ and $\overline{y}$ have finite limits as $t \to \infty$. This completes the assertion.

Remark 3.8 Suppose that $P(t_0,\infty)<\infty$ and $R(t_0,\infty)<\infty$. Then, as a result of this, we have $I_1<\infty$. So Theorem 3.7 also holds for $P(t_0,\infty)<\infty$ and $R(t_0,\infty)<\infty$.

Exercise 3.9 Prove Remark 3.8.

3.1.3 The case $P(t_0,\infty)<\infty$ and $R(t_0,\infty)=\infty$

We present the nonoscillation criteria in N^+ under the case $P(t_0,\infty)<\infty$ and $R(t_0,\infty)=\infty$ in this subsection. Therefore, we have the following lemma.

Lemma 3.10 *Suppose that $R(t_0,\infty)=\infty$. Then any nonoscillatory solution in N^+ belongs to $N^+_{F,\infty}$ or $N^+_{\infty,\infty}$, i.e., $N^+_{F,F}=N^+_{\infty,F}=\emptyset$.*

Exercise 3.11 Prove Lemma 3.10.

The following theorem shows us the nonexistence of nonoscillatory solutions in $N^+_{F,\infty}$. We skip the proof of the following theorem, since it is very similar to the proof of Theorem 3.5.

Theorem 3.12 $N^+_{F,\infty}\neq\emptyset$ *if and only if $I_1<\infty$ for all $k_1>0$.*

3.1.4 Examples

Examples are great ways to see that theoretical claims actually work. Therefore, we provide two examples about the existence of nonoscillatory solutions of system (1). But before the examples, we need the following proposition because our examples consist of scattered points.

Proposition 1 ([5], Theorem 1.79) *Let $a,b\in\mathbb{T}$ and $h\in C_{rd}$. If $[a,b]$ consists of only isolated points, then*

$$\int_a^b h(t)\Delta t=\sum_{t\in[a,b)_{\mathbb{T}}}\mu(t)h(t).$$

Example 3.13 *Let $\mathbb{T}=2^{\mathbb{N}_0}$. Consider*

$$\begin{cases}\Delta_q x(t)=\left(\dfrac{t}{2t-1}\right)^{\frac{1}{61}}(y(t))^{\frac{1}{61}}\\[2ex] \Delta_q y(t)=\dfrac{1}{2t^{\frac{13}{5}}}(x(t))^{\frac{3}{5}},\end{cases}\tag{9}$$

where Δ_q is known as a q-derivative and defined as $\Delta_q h(t)=\frac{h(\sigma(t))-h(t)}{\mu(t)}$, where $\mu(t)=t$, $\sigma(t)=2t$, and $t=2^n$, (see [5]). In this example, it is shown that we have a nonoscillatory solution in $N^+_{\infty,F}$ to highlight Theorem 3.5. Therefore, we need that $P(t_0,\infty)$ is divergent and $R(t_0,\infty)$ is convergent. Indeed, by Proposition 1, we have

$$P(1,T)=\int_1^T p(t)\Delta t=\sum_{t\in[1,T)_{2^{\mathbb{N}_0}}}\left(\frac{t}{2t-1}\right)^{\frac{1}{61}}\cdot t.$$

Hence, we have $P(1,\infty)=\infty$ as T tends to infinity. Note that we use the limit divergence test to show the divergence of $P(1,\infty)$. Next, we continue with the convergence of $R(1,\infty)$. To do that, we note

$$R(1,T)=\int_1^T r(t)\Delta t=\sum_{t\in[1,T)_{2^{\mathbb{N}_0}}}\frac{1}{2t^{\frac{13}{5}}}\cdot t.$$

As $T \to \infty$*, we have*

$$\sum_{n=0}^{\infty} \left(\frac{1}{2 \cdot 2^n}\right)^{\frac{8}{5}} < \infty$$

by the geometric series, i.e., $R(1,\infty) < \infty$. Finally, we have to show $I_2 < \infty$. Let $k_2 = 1$. Then we get

$$\int_1^T r(t) g\left(k_2 \int_1^t p(s)\Delta s\right)\Delta t = \int_1^T \frac{1}{2t^{\frac{13}{5}}}\left(\sum_{s\in[1,t)_{2^{\mathbb{N}_0}}} \left(\frac{s}{2s-1}\right)^{\frac{1}{61}} \cdot s\right)^{\frac{3}{5}} \Delta t$$

$$\leq \int_1^T \frac{1}{2t^{\frac{13}{5}}}\left(\sum_{s\in[1,t)_{2^{\mathbb{N}_0}}} s^{\frac{62}{61}}\right)^{\frac{3}{5}} \Delta t \leq \int_1^T \frac{1}{2t^{\frac{13}{5}}} \cdot t^{\frac{62}{105}} \Delta t = \sum_{t\in[1,T)_{2^{\mathbb{N}_0}}} \frac{1}{t^{\frac{208}{105}}}.$$

So as $t \to \infty$*, we have*

$$\sum_{n=0}^{\infty} \left(\frac{1}{2^n}\right)^{\frac{208}{105}} < \infty$$

by the ratio test. Therefore, $I_2 < \infty$ by the comparison test. One can also show that $\left(t, 2 - \frac{1}{t}\right)$ is a solution of system (9) such that $x(t) \to \infty$ and $y(t) \to 2$ as $t \to \infty$, i.e., $N^+_{\infty,F} \neq \emptyset$ by Theorem 3.5

Example 3.14 *Let* $\mathbb{T} = \left\{\frac{n}{2} : n \in \mathbb{N}_0\right\}, f(z) = z^{\frac{1}{3}}, g(z) = z^{\frac{1}{5}}, p(t) = \frac{\sqrt{2}(\sqrt{2}-1)}{2^{\frac{2t}{3}}(3\cdot 2^t - 1)^{\frac{1}{3}}}$, $r(t) = \frac{\sqrt{2}(\sqrt{2}-1)}{2^{\frac{4t}{5}}(2\cdot 2^t - 1)^{\frac{1}{5}}}$*, and* $t = \frac{n}{2}$ *in system* (1)*. We show that there exists a nonoscillatory solution in* $N^+_{F,F}$*. So by Theorem 3.7, we need to show* $P(t_0,\infty) < \infty$ *and* $R(t_0,\infty) < \infty$ *and* $I_1 < \infty$*. Proposition 1 gives us*

$$\int_0^T p(t)\Delta t = \sum_{t\in[0,T)_{\mathbb{T}}} \frac{\sqrt{2}(\sqrt{2}-1)}{2^{\frac{2t}{3}}(3\cdot 2^t - 1)^{\frac{1}{3}}} \cdot \frac{1}{2} \leq \sum_{t\in[0,T)_{\mathbb{T}}} \frac{1}{2^{\frac{2t}{3}}}.$$

So as $T \to \infty$*, we have*

$$\sum_{n=0}^{\infty} \frac{1}{2^{\frac{n}{3}}} < \infty$$

by the geometric series, i.e., $P(t_0,\infty) < \infty$. Also

$$\int_0^T r(t)\Delta t = \sum_{t\in[0,T)_{\mathbb{T}}} \frac{\sqrt{2}(\sqrt{2}-1)}{2^{\frac{4t}{5}}(2\cdot 2^t - 1)^{\frac{1}{5}}} \cdot \frac{1}{2} \leq \sum_{t\in[0,T)_{\mathbb{T}}} \frac{1}{2^{\frac{4t}{5}}}.$$

Hence, we have

$$\sum_{n=0}^{\infty} \frac{1}{2^{\frac{2n}{5}}} < \infty$$

as $T \to \infty$*. Note also that* $I_1 < \infty$ *if* $P(t_0,\infty) < \infty$ *and* $R(t_0,\infty) < \infty$ *(see Remark* (8)*). It can be confirmed that* $\left(2 - \frac{1}{2^t}, 3 - \frac{1}{2^t}\right)$ *is a nonoscillatory solution of*

$$\begin{cases} x^{\Delta}(t) = \dfrac{\sqrt{2}(\sqrt{2}-1)}{2^{\frac{2t}{3}}(3\cdot 2^{t}-1)^{\frac{1}{3}}}(y(t))^{\frac{1}{3}} \\ y^{\Delta}(t) = \dfrac{\sqrt{2}(\sqrt{2}-1)}{2^{\frac{4t}{5}}(2\cdot 2^{t}-1)^{\frac{1}{5}}}(x(t))^{\frac{1}{5}} \end{cases}$$

such that $x(t) \to 2$ and $y(t) \to 3$ as $t \to \infty$, i.e., $N^{+}_{F,F} \neq \emptyset$ by Theorem 3.7.

3.2 Existence of nonoscillatory solutions of (1) in N^{-}

Suppose that (x, y) is a nonoscillatory solution of system (1) such that $x > 0$ eventually. Then by the first and second equations of system (1) and the similar discussion as in Section 3.1, we obtain the following subclasses of N^{-}.

$$N^{-}_{F,F} = \left\{(x,y) \in N^{-} : \lim_{t\to\infty} x(t) = c, \quad \lim_{t\to\infty} y(t) = -d\right\},$$

$$N^{-}_{F,0} = \left\{(x,y) \in N^{-} : \lim_{t\to\infty} x(t) = c, \quad \lim_{t\to\infty} y(t) = 0\right\},$$

$$N^{-}_{0,F} = \left\{(x,y) \in N^{-} : \lim_{t\to\infty} x(t) = 0, \quad \lim_{t\to\infty} y(t) = -d\right\},$$

$$N^{-}_{0,0} = \left\{(x,y) \in N^{-} : \lim_{t\to 0} x(t) = 0, \quad \lim_{t\to 0} y(t) = 0\right\}.$$

This section presents us the existence and nonexistence of nonoscillatory solutions of system (1) under the monotonicity condition on f and g.

Theorem 3.15 *Let $R(t_0, \infty) < \infty$. Then there exists a nonoscillatory solution in $N^{-}_{F,F} \neq \emptyset$ if and only if $I_3 < \infty$ for all $k_3 < 0$ and $k_4 > 0$.*

Proof. Suppose $N^{-}_{F,F} \neq \emptyset$. Then there exists a solution $(x, y) \in N^{-}_{F,F}$ such that $x > 0, y < 0, x(t) \to c_1$, and $y(t) \to -d_1$ as $t \to \infty$ for $0 < c_1 < \infty$ and $0 < d_1 < \infty$. By integrating the second equation of system (1) from t to ∞, we obtain

$$\begin{aligned} y(t) &= y(\infty) - \int_t^{\infty} r(s)g(x(s))\Delta s \\ &\leq -d_1 - k_4 \int_t^{\infty} r(s)\Delta s, \quad \text{where} \quad k_4 = g(c_1). \end{aligned} \tag{10}$$

Integrating the first equation from t_1 to t, using (10) and the fact that x is bounded yield us

$$\begin{aligned} c_1 \leq x(t) &= x(t_1) + \textstyle\int_{t_1}^{t} p(s)f(y(s))\Delta s \\ &\leq x(t_1) + \textstyle\int_{t_1}^{t} p(s)f\left(-d_1 - k_4 \int_s^{\infty} r(u)\Delta u\right)\Delta s \leq x(t_1), \quad t \geq t_1. \end{aligned}$$

Therefore, it implies $I_3 < \infty$ as $t \to \infty$, where $-d_1 = k_3$.

Conversely, suppose that $I_3 < \infty$. Then there exist $t_1 \geq t_0$ and $k_3 < 0, k_4 > 0$ such that

$$\int_{t_0}^{\infty} p(t)f\left(k_3 - k_4 \int_t^{\infty} r(s)\Delta s\right)\Delta t > \frac{-1}{2} \tag{11}$$

where $k_4 = g\left(\frac{3}{2}\right)$. Let C_B be the set of all continuous and bounded real-valued functions $x(t)$ on $[t_1, \infty)_{\mathbb{T}}$ with the supremum norm $\sup_{t \geq t_1} |x(t)|$. Observe that C_B is a Banach space (see [38]). Suppose that B is a subset of C_B such that

$$B := \left\{ x(t) \in C_B : 1 \leq x(t) \leq \frac{3}{2}, \quad t \geq t_1 \right\}.$$

We have that B meets the assumptions of Theorem 2.9. Suppose also that $F : B \to B$ is an operator such that

$$(Fx)(t) = 1 - \int_t^\infty a(s) f\left(k_3 - \int_s^\infty b(u) g(x(u)) \Delta u \right) \Delta s \tag{12}$$

First, we need to show F is a mapping into itself, i.e., $F : B \to B$. Indeed,

$$1 \leq (Fx)(t) \leq 1 - \int_t^\infty a(s) f\left(k_3 - g\left(\frac{3}{2}\right) \int_{t_1}^s b(u) \Delta u \right) \Delta s \leq \frac{3}{2}$$

because $x \in B$ and (5) hold. Next, let us verify that F is continuous on B. In order to do that, let x_n be a sequence in B such that $x_n \to x,$ where $x \in B = \overline{B}$. Then

$$\begin{aligned} &|(Fx_n)(t) - (Fx)(t)| \\ &\leq \int_t^\infty p(s) \left| f\left(k_3 - \int_s^\infty r(u) g(x_n(u)) \Delta u \right) - f\left(k_3 - \int_s^\infty r(u) g(x(u)) \Delta u \right) \right| \Delta s. \end{aligned}$$

Therefore, the continuity of f and g and the Lebesgue dominated convergence theorem gives us $Fx_n \to Fx$ as $n \to \infty$, which implies F is continuous on B. Finally, we prove that FY is equibounded and equicontinuous, i.e., relatively compact. Because

$$\begin{aligned} 0 < -(Fx)^\Delta(t) &= -p(t) f\left(k_3 - \int_t^\infty r(u) g(x(u)) \Delta u \right) \\ &\leq -p(t) f\left(k_3 - k_4 \int_{t_1}^t r(u) \Delta u \right) < \infty, \end{aligned}$$

we have that Fx is relatively compact. Hence, Theorem 2.9 implies that there exists $\overline{x} \in B$ such that $\overline{x} = F\overline{x}$. Thus, we have $\overline{x} > 0$ eventually and $\overline{x}(t) \to 1$ as $t \to \infty$. Also

$$\overline{x}^\Delta(t) = (F\overline{x})^\Delta(t) = p(t) f\left(k_3 - \int_t^\infty r(u) g(\overline{x}(u)) \Delta u \right) \quad t \geq t_1.$$

Letting

$$\overline{y}(t) = k_3 - \int_t^\infty r(u) g(\overline{x}(u)) \Delta u < 0, \quad t \geq t_1 \tag{13}$$

and taking the derivative of (13) give $\overline{y}^\Delta(t) = b(t) g(\overline{x}(t))$. So, we conclude that $(\overline{x}, \overline{y})$ is a nonoscillatory solution of system (1). Finally, taking the limit of Eq. (13) results in $\overline{y}(t) \to k_3 < 0$. Therefore, we get $N_{F,F}^- \neq \emptyset$.

Theorem 3.16 *Suppose* $P(t_0, \infty) < \infty$. $N_{0,F}^{-} \neq \emptyset$ *if and only if* $I_4 < \infty$ *for* $k_5 > 0$.

Exercise 3.17 Prove Theorem 3.16.

Theorem 3.18 *Suppose* $P(t_0, \infty) < \infty$. $N_{0,0}^{-} \neq \emptyset$ *if* $I_3 < \infty$ *and* $I_4 = \infty$ *for all* $k_3 = 0, k_4 < 0$ *and* $k_5 > 0$, *provided* f *is odd.*

Proof. Suppose that $I_3 < \infty$, and $I_4 = \infty$. Then there exists $t_1 \geq t_0$ such that

$$\int_{t_1}^{\infty} p(s) f\left(-k_4 \int_s^{\infty} r(u)\Delta u\right)\Delta s < 1$$

and

$$\int_{t_1}^{\infty} r(s) g\left(k_5 \int_s^{\infty} p(u)\Delta u\right)\Delta s > \frac{1}{2}$$

for $t \geq t_1$, $k_4 = -g(1)$. Let X be the space that is claimed as in the proof of Theorem 3.7. Let Y be a subset of X and given by

$$Y := \left\{x \in X : c_1 \int_t^{\infty} a(s)\Delta s \leq x(t) \leq 1 \quad t \geq t_1\right\},$$

where $c_1 = f\left(\frac{1}{2}\right)$. Define an operator $T : Y \to X$ such that

$$(Tx)(t) = \int_t^{\infty} p(s) f\left(\int_s^{\infty} r(u) g(x(u))\Delta u\right)\Delta t, \quad t \geq t_1.$$

One can show that $(Y, \leq)$ is a complete lattice and T is an increasing mapping such that $T : Y \to Y$. As a matter of fact,

$$(Tx)(t) \leq \int_t^{\infty} p(s) f\left(g(1) \int_s^{\infty} r(u)\Delta u\right)\Delta t \leq 1, \quad t \geq t_1$$

and

$$(Tx)(t) \geq \int_t^{\infty} p(s) f\left(\int_s^{\infty} r(u) g\left(c_1 \int_u^{\infty} p(v)\Delta v\right)\Delta u\right)\Delta s$$
$$\geq f\left(\frac{1}{2}\right) \int_t^{\infty} p(s)\Delta s,$$

where $c_1 = k_5$, i.e., $T : Y \to Y$. Then by Theorem 2.10, there exists a function $\overline{x} \in Y$ such that $\overline{x} = T\overline{x}$. By taking the derivative of $T\overline{x}$ and using the fact that f is odd, we have

$$(T\overline{x})^{\Delta}(t) = p(t) f\left(-\int_t^{\infty} r(u) g(\overline{x}(u))\Delta u\right), \quad t \geq t_1.$$

Setting

$$\overline{y}(t) = -\int_t^{\infty} r(u) g(\overline{x}(u))\Delta u$$

yields $\overline{y}^{\Delta}(t) = b(t)g(\overline{x}(t))$, and $(\overline{x}, \overline{y})$ is a solution of system (1) in $N^{-}_{0,0}$, i.e., $\overline{x}$ and $\overline{y}$ both tend to zero.

Theorem 3.19 *Suppose $R(t_0, \infty) < \infty$. $N^{-}_{F,0} \neq \emptyset$ if and only if $I_3 < \infty$, where $k_3 = 0$ and $k_4 > 0$.*

Exercise 3.20 Prove Theorem 3.19. Hint: Use Theorem 2.10 with the operator

$$(Fx)(t) = \frac{1}{2} - \int_t^{\infty} a(s) f\left(- \int_s^{\infty} b(u) g(x(u)) \Delta u \right) \Delta t, \quad t \geq t_1.$$

Examples make results clearer and give more information to readers. Therefore, we give the following example to validate our claims. The beauty of our example is that we do not only show the theorem holds but also find the explicit solutions, which might be very hard for some nonlinear systems.

Example 3.21 *Consider $\mathbb{T} = \mathbb{N}_0^2 = \{n^2 : n \in \mathbb{N}_0\}$ with the system*

$$\begin{cases} x^{\Delta}(t) = \dfrac{1}{t^{\frac{1}{3}}\left(\sqrt{t}+1\right)^2 (t^2+1)^{\frac{1}{3}}} (y(t))^{\frac{1}{3}} \\ y^{\Delta}(t) = \dfrac{\left(\sqrt{t}+1\right)^4 - t^2}{t^{\frac{9}{5}}\left(\sqrt{t}+1\right)^4 \left(1+2\sqrt{t}\right)} (x(t))^{\frac{1}{5}}, \end{cases} \tag{14}$$

where $f^{\Delta}(t) = \frac{f(\sigma(t)) - f(t)}{\mu(t)}$ for $\sigma(t) = \left(\sqrt{t}+1\right)^2$ and $\mu(t) = 1 + 2\sqrt{t}$ (see [5]). *First, let us show $P(t_0, \infty) < \infty$, where $t_0 \geq 1$.*

$$\int_1^T p(t) \Delta t = \sum_{t \in [1,T)_{\mathbb{N}_0^2}} \frac{1}{t^{\frac{1}{3}}\left(\sqrt{t}+1\right)^2 (t^2+1)^{\frac{1}{3}}} \cdot \left(1+2\sqrt{t}\right) \leq \sum_{t \in [1,T)_{\mathbb{N}_0^2}} \frac{1+2\sqrt{t}}{t^2}.$$

Since $t = n^2$, as $T \to \infty$, we have

$$\sum_{n=1}^{\infty} \frac{1+2n}{n^4} < \infty$$

by the geometric series. Therefore, $P(1, \infty) < \infty$ by the comparison test. Next, we show $I_4 < \infty$. Since $P(1, \infty) < \infty$, we have $\int_t^{\infty} p(s) \Delta s < \alpha$ for $t \geq 1$ and $0 < \alpha < \infty$. Hence,

$$\int_1^T r(t) g\left(\int_t^{\infty} p(s) \Delta s \right) \Delta t \leq \alpha \int_1^T r(t) \Delta t$$

$$= \alpha \sum_{t \in [1,T)_{\mathbb{N}_0^2}} \frac{\left(\sqrt{t}+1\right)^4 - t^2}{t^{\frac{9}{5}}\left(\sqrt{t}+1\right)^4 \left(1+2\sqrt{t}\right)} \cdot \left(1+2\sqrt{t}\right)$$

$$\leq \alpha \sum_{t \in [1,T)_{\mathbb{N}_0^2}} \frac{1}{t^{\frac{9}{5}}}.$$

So as T tends to infinity, we get

$$\sum_{n=1}^{\infty} \frac{1}{n^{\frac{18}{5}}} < \infty,$$

i.e., $I_2 < \infty$. *Also, note that* $\left(\frac{1}{t}, -1-\frac{1}{t^2}\right)$ *is a solution of system* (14) *in* N^- *such that* x *tends to zero, while* y *tends to* -1, *i.e.,* $N^-_{0,F} \neq \emptyset$.

4. Oscillation of a two-dimensional time-scale systems

Motivated by [39], this section deals with the system

$$\begin{cases} x^{\Delta}(t) = a(t)f(y(t)) \\ y^{\Delta}(t) = -b(t)g(x(t)) + c(t), \end{cases} \tag{15}$$

where $a, b \in C_{rd}\left([t_0,\infty)_{\mathbb{T}}, \mathbb{R}^+\right)$, $c \in C_{rd}\left([t_0,\infty)_{\mathbb{T}}, \mathbb{R}\right)$ and functions f g have the same characteristics as in system (1) and g is continuously differentiable. Note that we can rewrite system (15) as a non-homogenous dynamic equations on time scales and putting σ on x inside the function g. Therefore, we have the following dynamic equation

$$\left(a(t)x^{\Delta}(t)\right)^{\Delta} + b(t)g(x^{\sigma}(t)) = c(t) \tag{16}$$

and systems of dynamical equations

$$\begin{cases} x^{\Delta}(t) = a(t)f(y(t)) \\ y^{\Delta}(t) = -b(t)g(x^{\sigma}(t)) + c(t). \end{cases} \tag{17}$$

Oscillation criteria for Eq. (16), system (17), and other similar versions of (15) and (17) are investigated in [39–42]. A solution (x, y) of system (15) is called oscillatory if x and y have arbitrarily large zeros. System (15) is called oscillatory if all solutions are oscillatory.

Before giving the main results, we present some propositions so that we can use them in our theoretical claims (see [43], Theorem 4.2 (comparison theorem) and [5], Theorem 1.90).

Proposition 2 *Let* z_1 *be a function from* $\mathbb{T}$ *to* $\mathbb{R}$ *and* v *be a nondecreasing function from* $\mathbb{R}$ *to* $\mathbb{R}$ *such that* $v \circ z_1$ *is rd-continuous. Suppose also that* $p \geq 0$ *is rd-continuous and* $\alpha \in \mathbb{R}$. *Then*

$$z_1(t) \leq \alpha + \int_{t_0}^{t} p(\tau)v(z_1(\tau))\Delta\tau, \quad t \geq t_0$$

implies $z_1(t) \leq z_2(t)$, where z_2 solves the initial value problem

$$z_2^{\Delta}(t) = p(t)v(z_2(t)), \quad z_2(t_0) = z_{20} > \alpha.$$

Proposition 3 (chain rule). ([5], Theorem 1.90) *Let* $h_1 : \mathbb{R} \to \mathbb{R}$ *be continuously differentiable and suppose* $h_2 : \mathbb{T} \to \mathbb{R}$ *is delta differentiable. Then* $h_1 \circ h_2 : \mathbb{T} \to \mathbb{R}$ *is delta differentiable, and the formula*

$$(h_1 \circ h_2)^{\Delta}(t) = \left\{ \int_0^1 h_{1'}\left(h_2(t) + h\mu(t)h_2^{\Delta}(t)\right)dh \right\} h_2^{\Delta}(t)$$

holds.

For simplicity, set

$$A(t,s)=\int_t^s a(u)\Delta u,\quad B(t,s)=\int_t^s b(u)\Delta u,$$

$$C(t,s)=\int_t^s |c(u)|\Delta u,\quad D(t,s)=\int_t^s \left(b(u)-\frac{c(u)}{g(x(u))}\right)\Delta u,$$

$$Y(t,s)=\int_t^s \frac{y^\sigma(u)x^\Delta(u)\int_0^1 [g'(x(u)+h\mu(u)x^\Delta(u))dh]}{g(x(u))g(x^\sigma(u))}\Delta u.$$

Next, note that if (x,y) is a nonoscillatory solution of system (15), then one can easily prove that x is also nonoscillatory. This result was shown by Anderson in [37] when $c(t)\equiv 0$. Because the proof when $c(t)\not\equiv 0$ is very similar to the proof of the case $c(t)\equiv 0$, we leave it to the readers.

Lemma 4.1 *Suppose that* (x,y) *is a nonoscillatory solution of system* (15) *and* $t_1,t_2\in\mathbb{T}$. *If there exists a constant* $K>0$ *such that*

$$H(t)\geq K,\quad t\geq t_2, \tag{18}$$

where H is defined as

$$H(t)=-\frac{y(t_1)}{g(x(t_1))}+D(t_1,t)+Y(t_1,t_2), \tag{19}$$

then $y(t)\leq -Kg(x(t_2)),\quad t\geq t_2.$

Proof. Suppose that (x,y) is a nonoscillatory solution of system (15). Then, we have that x is also nonoscillatory. Without loss of generality, assume that $x(t)>0$ for $t\geq t_1\geq t_0$, where $t_1,t_0\in\mathbb{T}$. Integrating the second equation of system (15) from t_1 to t and Theorem 2.8 (v.) gives us

$$\int_{t_1}^t b(s)\Delta s=\frac{y(t_1)}{g(x(t_1))}-\frac{y(t)}{g(x(t))}+\int_{t_1}^t\left(\frac{1}{g(x(s))}\right)^\Delta y^\sigma(s)\Delta s+\int_{t_1}^t\frac{c(s)}{g(x(s))}\Delta s. \tag{20}$$

By applying Theorem 2.3 (iv) and Proposition 3 to Eq. (20), we have

$$\int_{t_1}^t b(s)\Delta s=\frac{y(t_1)}{g(x(t_1))}-\frac{y(t)}{g(x(t))}+\int_{t_1}^t\frac{c(s)}{g(x(s))}\Delta s-Y(t_1,t),\quad t\geq t_1. \tag{21}$$

Rewriting Eq. (21) gives us

$$-\frac{y(t)}{g(x(t))}=D(t_1,t)-\frac{y(t_1)}{g(x(t_1))}+Y(t_1,t),\quad t\geq t_1. \tag{22}$$

Now by using (18) and (19), we get

$$-\frac{y(t)}{g(x(t))}\geq K+Y(t_2,t),\quad t\geq t_2\geq t_1. \tag{23}$$

Note that $y(t)<0$ and $x^\Delta(t)<0$ for $t\geq t_2$ since $y(t)x^\Delta(t)=a(s)y(s)f(y(s))>0$. Otherwise, we would have $\frac{-y(t)}{g(x(t))}>0$, which is a contradiction. Let

$$\frac{-v(t)}{g(x(t))}=K+Y(t_2,t),\quad t\geq t_2. \tag{24}$$

So one can obtain

$$\left(\frac{-v(t)}{g(x(t))}\right)^{\Delta}=\frac{y^{\sigma}(t)x^{\Delta}(t)\int_0^1[g'(x(t)+h\mu(t)x^{\Delta}(t))dh]}{g(x(t))g(x^{\sigma}(t))}>0,\quad t\geq t_2. \tag{25}$$

Because $x(t)$ is a positive and $v(t)$ is a negative function for $t\geq t_2$, we have $\frac{-y(t)}{g(x(t))}\geq\frac{-v(t)}{g(x(t))}$, i.e., $y(t)\leq v(t)<0$ for $t\geq t_2$. Therefore, we have by (25) that

$$\left(\frac{-v(t)}{g(x(t))}\right)^{\Delta}\geq\frac{v^{\sigma}(t)x^{\Delta}(t)\int_0^1[g'(x(t)+h\mu(t)x^{\Delta}(t))dh]}{g(x(t))g(x^{\sigma}(t))}>0,\quad t\geq t_2$$

since $v(t)<0$ and $x^{\Delta}(t)<0$ for $t\geq t_2$. By setting

$$\frac{w(t)}{g(x(t))}=K-\int_{t_2}^{t}\frac{w^{\sigma}(s)x^{\Delta}(s)\int_0^1[g'(x(s)+h\mu(s)x^{\Delta}(s))dh]}{g(x(s))g(x^{\sigma}(s))}\Delta s \tag{26}$$

and using (24), we have $\frac{-v(t_2)}{g(x(t_2))}=K=\frac{w(t_2)}{g(x(t_2))}$. Then, setting $z_1=\frac{v(t)}{g(x(t))}$, $z_2=\frac{-w(t)}{g(x(t))}$, $h(u)=\frac{u^{\sigma}(t)}{g(x(t))}$ in Proposition 2, it follows $v(t)\leq-w(t)$, which implies $y(t)\leq-w(t)$, $t\geq t_2$. Note also by Theorem 2.3 (iv) and Proposition 3 that

$$\left(\frac{w(t)}{g(x(t))}\right)^{\Delta}=\frac{w^{\Delta}(t)}{g(x^{\sigma}(t))}-\frac{w^{\sigma}(t)x^{\Delta}(t)\int_0^1 g'(x(t)+h\mu(t)x^{\Delta}(t))dh}{g(x(t))g(x^{\sigma}(t))},\quad t\geq t_2. \tag{27}$$

Taking the derivative of (26) and comparing the resulting equation with (27) yield us

$$\frac{w^{\Delta}(t)}{g(x^{\sigma}(t))}=0,\quad i.e., w^{\Delta}(t)=0,\quad t\geq t_2.$$

Therefore, we have

$$w(t_2)=K\cdot g(x(t_2))=w(t),\quad i.e., y(t)\leq-w(t)=-K\cdot g(x(t_2)).$$

So the proof is completed.

4.1 Results for oscillation

After giving the preliminaries in the previous section, it is presented the conditions for oscillatory solutions in this section.

Theorem 4.2 *Let* $A(t_0,\infty)=\infty$, $B(t_0,\infty)<\infty$, *and* $C(t_0,\infty)<\infty$. *Assume*

$$f(u)f(v)\leq f(uv)\leq-f(u)f(-v) \tag{28}$$

and

$$\int_{t_0}^{\infty}\frac{x^{\Delta}(s)}{f(g(x(s)))}\Delta s<\infty. \tag{29}$$

Then system (15) *is oscillatory if*

$$\int_{t_0}^{\infty}a(t)f(B(t,\infty)-k\cdot C(t,\infty))\Delta t=\infty \tag{30}$$

for $k \neq 0$.

Proof. Suppose that system (15) has a nonoscillatory solution (x,y) such that $x>0$ eventually. Then there exist $t_1 \geq t_0$ and a constant k_6 such that $g(x(t)) \geq k_6$ for $t \geq t_1$ by the monotonicity of g. Then by Eq. (22), we have

$$\frac{y(t)}{g(x(t))} = \frac{y(t_1)}{g(x(t_1))} - D(t_1,t) - Y(t_1,t), \quad t \geq t_1. \tag{31}$$

Note that $Y(t_1,t) < \infty$. Otherwise, we have a contradiction to the fact that $x(t)>0$ for $t \geq t_1$ since $A(t_0,\infty)=\infty$. Equality (31) can be rewritten as

$$\frac{y(t)}{g(x(t))} = \gamma + D(t,\infty) + Y(t,\infty), \tag{32}$$

where $\gamma = \frac{y(t_1)}{g(x(t_1))} - D(t_1,\infty) - Y(t_1,\infty), \quad t \geq t_1$. It can be shown that $\gamma \geq 0$. Otherwise, we can choose a large t_2 such that $B(t,\infty) \leq -\gamma$, $Y(t_2,\infty) \leq \frac{-\gamma}{4}$, and $\left|\int_t^\infty \frac{c(s)}{g(x(s))}\Delta s\right| \leq \frac{-\gamma}{4}$ for $t \geq t_2$. Then $H(t) \geq \frac{-\gamma}{4} > 0$ for $t \geq t_2$. Then by setting $K = \frac{-\gamma}{4}$ in Lemma 4.1 found, we have $y(t) \leq -Kg(x(t_2))$ for $t \geq t_2$. Integrating the first equation of system (15) from t_2 to ∞ and the monotonicity of f yields us

$$x(t) \leq x(t_2) + f(-Kg(x(t_2))) \int_{t_2}^t a(s)\Delta s, \quad t \geq t_2.$$

So as $t \to \infty$, we have a contradiction to $x>0$ eventually. Therefore $\gamma \geq 0$. Then by Eq. (32), we have

$$y(t) \geq g(x(t)) \left[\int_t^\infty b(s)\Delta s - \frac{1}{k_6}\int_t^\infty |c(s)|\Delta s\right], \quad t \geq t_2.$$

By the first equation of system (15), the monotonicity of f and Eq. (28), we have

$$x^\Delta(t) \geq a(t) f(g(x(t))) f\left(\int_t^\infty b(s)\Delta s - \frac{1}{k_6}\int_t^\infty |c(s)|\Delta s\right), \quad t \geq t_2. \tag{33}$$

Then by Eqs. (33) and (29), we have

$$\int_{t_2}^t a(s) f\left(\int_s^\infty b(u)\Delta u - k\int_s^\infty |c(u)|\Delta u\right) \leq \int_{t_2}^t \frac{x^\Delta(s)}{f(g(x(s)))}\Delta s < \infty$$

where $k = \frac{1}{k_6}$. But as $t \to \infty$, this contradicts to Eq. (30). The proof is completed.

Theorem 4.3 *System (15) is oscillatory if* $A(t_0,\infty) = B(t_0,\infty) = \infty$ *and* $C(t_0,\infty) < \infty$.

Proof. We use the method of contradiction to prove the theorem. Thus, assume there is a nonoscillatory solution (x,y) of system (15) such that the component function x is eventually positive. Because g is nondecreasing, we have that there exist $t_1 \geq t_0$ and $k_7 > 0$ such that $g(x(t)) \geq k_7$ for $t \geq t_1$. Then since $C(t_0,\infty) < \infty$, we have that there exists $0 < k_8 < \infty$ such that

$$\left|\int_{t_1}^t \frac{c(s)}{g(x(s))}\Delta s\right| \leq \frac{1}{k_7}\int_{t_1}^t |c(s)|\Delta s < k_8, \quad t \geq t_1. \tag{34}$$

The first equation of system (15), and the monotonicity of g give us that there exist $K > 0$ and $t_2 \geq t_1$ so large that

$$x^{\Delta}(t) \leq a(t)f(-Kg(x(t_2))), \quad t \geq t_2. \tag{35}$$

Integrating (35) from t_2 to t yields

$$x(t) \leq x(t_2) + k_9 \int_{t_2}^{t} a(s)\Delta s, \quad \text{where} \quad k_9 = f(-Kg(x(t_2))) < 0, \quad t \geq t_2.$$

As $t \to \infty$, we have a contradiction to $x(t) > 0$ for $t \geq t_2$. This proves the assertion.

Finally, an example is provided to highlight Theorem 4.3 by finding the explicit solution of the dynamical system.

Example 4.4 *Consider the time scale* $\mathbb{T} = 5\mathbb{Z}^+$ *with* $a(t) = \frac{(t+4)^{\frac{1}{3}}(2t+7)}{5(t+1)^{\frac{2}{3}}(t+6)}$, $b(t) = \frac{t^5+t^4+t^3+t^2+t+1}{5(t+1)(t+4)(t+6)(t+9)}$, $f(z) = z^{\frac{1}{3}}, g(z) = z^3$, $c(t) = \frac{(-1)^{3t}\left(-3t^5-27t^4-125t^3-237t^2-195t-59\right)}{5(t+1)^4(t+4)(t+6)(t+9)}$, *and* $t = 5n$, *where* $n \in \mathbb{N}$ *in system* (15). *We show that* $A(t_0, \infty) = \infty$, $B(t_0, \infty) = \infty$, *and* $C(t_0, \infty) < \infty$. *Indeed,*

$$A(5, T) = \int_5^T \frac{(t+4)^{\frac{1}{3}}(2t+7)}{5(t+1)^{\frac{2}{3}}(t+6)}\Delta t = \sum_{t \in [5,T)_{5\mathbb{Z}^-}} \frac{(t+4)^{\frac{1}{3}}(2t+7)}{(t+1)^{\frac{2}{3}}(t+6)}.$$

So as $T \to \infty$, *we have*

$$\sum_{n=1}^{\infty} \frac{(5n+4)^{\frac{1}{3}}(10n+7)}{(5n+1)^{\frac{2}{3}}(5n+6)} = \infty \quad \textit{by the limit comparison test. Therefore,} \quad A(5, \infty) = \infty.$$

Similarly,

$$\begin{aligned} B(5, T) &= \int_5^T \frac{t^5+t^4+t^3+t^2+t+1}{5(t+1)(t+4)(t+6)(t+9)}\Delta t = \sum_{t \in [5,T)_{5\mathbb{Z}^+}} \frac{t^5+t^4+t^3+t^2+t+1}{(t+1)(t+4)(t+6)(t+9)} \\ &\geq \sum_{t \in [5,T)_{5\mathbb{Z}^+}} \frac{t^5}{(t+1)(t+4)(t+6)(t+9)}. \end{aligned}$$

Taking the limit as $T \to \infty$ *gives us*

$$B(5, \infty) \geq 625 \cdot \sum_{n=1}^{\infty} \frac{n^5}{(5n+1)(5n+4)(5n+6)(5n+9)} = \infty$$

by the limit divergence test. Therefore, $B(5, \infty) = \infty$ *by the comparison test. Finally, we show* $C(t_0, \infty) < \infty$.

$$\begin{aligned} C(5, T) &= \sum_{t \in [5,T)_{5\mathbb{Z}^+}} \frac{3t^5+27t^4+125t^3+237t^2+195t+59}{(t+1)^4(t+4)(t+6)(t+9)} \\ &\leq \sum_{t \in [5,T)_{5\mathbb{Z}^+}} \frac{3}{t^2} + \frac{27}{t^3} + \frac{125}{t^5} + \frac{195}{t^6} + \frac{59}{t^7}. \end{aligned}$$

So as $T \to \infty$, *we have*

$$C(5,\infty) \leq \sum_{n=1}^{\infty} \frac{3}{n^2} + \frac{27}{n^3} + \frac{125}{n^5} + \frac{195}{n^6} + \frac{59}{n^7} < \infty$$

by the geometric series. One can also show that $\left(\frac{(-1)^{t+1}}{t+1}, \frac{(-1)^{3t}}{(t+1)(t+4)}\right)$ is an oscillatory solution of system

$$\begin{cases} x^{\Delta}(t) = \dfrac{(t+4)^{\frac{1}{3}}(2t+7)}{5(t+1)^{\frac{2}{3}}(t+6)} y^{\frac{1}{3}}(t) \\ y^{\Delta}(t) = -\dfrac{t^5+t^4+t^3+t^2+t+1}{5(t+1)(t+4)(t+6)(t+9)} x^3(t) + \dfrac{(-1)^{3t}(-3t^5-27t^4-125t^3-237t^2-195t-59)}{5(t+1)^4(t+4)(t+6)(t+9)}, \end{cases}$$

where we define $h^{\Delta}(t) = \frac{h(\sigma(t))-h(t)}{\mu(t)}$ for $\sigma(t) = t+5$ and $\mu(t) = 5$ (see [5]).

5. Conclusion

This chapter focuses on the oscillation/nonoscillation criteria of two-dimensional dynamical systems on time scales. We do not only show the oscillatory behaviors of such solutions but also guarantee the existence of such solutions, which might be challenging most of the time for nonlinear systems. In the first and second sections, we present some introductory parts to dynamical systems and basic calculus of the time-scale theory for the readers to comprehend the idea behind the time scales. In Section 3, we consider

$$\begin{cases} x^{\Delta}(t) = p(t)f(y(t)) \\ y^{\Delta}(t) = r(t)g(x(t)) \end{cases}$$

and investigate the nonoscillatory behavior of solutions under some certain circumstances. Recall that system (1) turns out to be a differential equation system

$$\begin{cases} x'(t) = p(t)f(y(t)) \\ y'(t) = r(t)g(x(t)) \end{cases}$$

when $\mathbb{T} = \mathbb{R}$. And the asymptotic behaviors of nonoscillatory solutions were presented by Li in [32]. Also when $\mathbb{T} = \mathbb{Z}$, system (1) is reduced to the difference equation system,

$$\begin{cases} \Delta x_n = p_n f(y_n) \\ \Delta y_n = r_n g(x_n), \end{cases}$$

and the existence of nonoscillatory solutions were investigated in [33]. Therefore, we unify the results for oscillation and nonoscillation theory, which was shown in $\mathbb{R}$ and $\mathbb{Z}$ and extends them in one comprehensive theory, which is called time-scale theory. These results were inspired from the book chapter written by Elvan Akın and Özkan Öztürk (see [29]). In that book chapter, it was considered a second-order dynamical system

$$\begin{cases} x^{\Delta}(t) = p(t)f(y(t)) \\ y^{\Delta}(t) = -r(t)g(x(t)) \end{cases} \tag{36}$$

and delay system

$$\begin{cases} x^{\Delta}(t) = p(t)f(y(t)) \\ y^{\Delta}(t) = -r(t)g(x((\tau(t))), \end{cases} \tag{37}$$

where τ is rd-continuous function such that $\tau(t) \leq t$ and $\tau(t) \to \infty$ as $t \to \infty$. When the latter systems were considered, because of the negative sign of the second equation of systems, the subclasses for N^+ an N^- would be totally different. So in [29], the existence of nonoscillatory solutions in different subclasses was shown. Another crucial thing on the results is that it is assumed that f must be an odd function for some main results. However, we do not have these strict conditions on our results. Another interesting observation for system (37) is that we lose some subclasses when we consider the delay in system (37). It is because of the setup fixed point theorem and the delay function τ. Therefore, this is a big disadvantage of delayed systems on time scales.

Akın and Öztürk also considered the system

$$\begin{cases} x^{\Delta}(t) = p(t)|y(t)|^{\alpha} \operatorname{sgn} y(t) \\ y^{\Delta}(t) = -r(t)|x(\sigma(t))|^{\beta} \operatorname{sgn} x^{\sigma}(t), \end{cases} \tag{38}$$

where $\alpha, \beta > 0$. System (38) is known as *Emden-Fowler* dynamical systems on time scales in the literature that has been mentioned in Section 1 with applications. Akın et al. [44, 45] showed the asymptotic behavior of nonoscillatory solutions by using α and β relations.

For example, system (38) turns out to be a system of first-order differential equation

$$\begin{cases} x'(t) = p(t)|y(t)|^{\alpha} \operatorname{sgn} y(t) \\ y'(t) = -r(t)|x(t)|^{\beta} \operatorname{sgn} x(t), \end{cases}$$

when the time scale $\mathbb{T} = \mathbb{R}$. On the other hand, system (38) ends up with the system of difference equations

$$\begin{cases} \Delta x_n = p_n |y_n|^{\alpha} \operatorname{sgn} y_n \\ \Delta y_n = -r_n |x_{n+1}|^{\beta} \operatorname{sgn} x_{n+1}, \end{cases}$$

when the time scale $\mathbb{T} = \mathbb{Z}$. For both cases, several contributions have been made by Zuzana et al. in [46] and [47], respectively.

Finally, we finish this section with the following tables, showing summaries about the existence of nonoscillatory solutions of system (1) in N^+ and N^- (**Tables 3** and **4**).

$N^+_{\infty,F}$	$\neq \emptyset$	$P(t_0,\infty) = \infty$ and $R(t_0,\infty) < \infty$	$I_2 < \infty$
$N^+_{F,F}$	$\neq \emptyset$	$P(t_0,\infty) < \infty$ and $R(t_0,\infty) < \infty$	$I_1 < \infty$
$N^+_{F,\infty}$	$\neq \emptyset$	$P(t_0,\infty) < \infty$ and $R(t_0,\infty) = \infty$	$I_1 < \infty$

Table 3.
Existence for (1) in N^+.

$N^-_{F,F}$	$\neq \emptyset$	$R(t_0,\infty) < \infty$	$I_3 < \infty$
$N^-_{0,F}$	$\neq \emptyset$	$P(t_0,\infty) < \infty$	$I_4 < \infty$
$N^-_{0,0}$	$\neq \emptyset$	$P(t_0,\infty) < \infty$	$I_3 < \infty$ and $I_4 = \infty$
$N^-_{F,0}$	$\neq \emptyset$	$R(t_0,\infty) < \infty$	$I_3 < \infty$

Table 4.
Existence for (1) in N^-.

A. Appendix

We give the following exercises to the interested readers that help them practicing the theoretical results. The examples are in q-calculus which takes too much attention recently. Recall from Example 3.13 that Δ_q is defined as

$$\Delta_q f(t) = \frac{f(tq) - f(t)}{(q-1)t}. \tag{39}$$

With the help of Eq. (39), we provide the following exercises.

Exercise 6.1 *Let* $\mathbb{T} = 2^{\mathbb{N}_0}$. *Consider the following system:*

$$\begin{cases} \Delta_q x(t) = \dfrac{1}{4t^2(1+t)^{\frac{1}{7}}}(y(t))^{\frac{1}{7}} \\ \Delta_q y(t) = \dfrac{2t}{4t-1}x(t) \end{cases} \tag{40}$$

and show that $\left(2 - \frac{1}{2t}, t+1\right)$ is a nonoscillatory solution of Eq. (40) in $N^+_{F,\infty} \neq \emptyset$ by checking the conditions given in Theorem 3.12 for $k_1 = 1$.

Exercise 6.2 *Let* $\mathbb{T} = q^{\mathbb{N}_0}, q > 1$. *Consider the following system:*

$$\begin{cases} \Delta x_q(t) = \dfrac{1}{qt^{\frac{8}{5}}(2t^2+1)^{\frac{1}{5}}}(y(t))^{\frac{1}{5}} \\ \Delta y_q(t) = \dfrac{q+1}{q^2t^2(t+1)}x(t), \end{cases} \tag{41}$$

where $\Delta h_q(t) = \frac{h(\sigma(t)) - h(t)}{\mu(t)}$ and show that there exists a nonoscillatory solution of system (41), given by $\left(1 + \frac{1}{t}, -2 - \frac{1}{t^2}\right)$, in $N^-_{F,F} \neq \emptyset$ by Theorem 3.15 for $k_3 = -1$ and $k_4 = 1$.

Author details

Ozkan Ozturk
Department of Mathematics, College of Engineering, American University of the Middle East, Eqaila, Kuwait

*Address all correspondence to: ozkan.ozturk@aum.edu.kw

References

[1] Hilger S. Ein Maßkettenkalkül mit Anwendung auf Zentrumsmannigfaltig keiten [thesis]. Universität Würzburg; 1988

[2] Agarwal RP. Difference Equations and Inequalities: Theory, Methods, and Applications. New York, NY, USA: Marcel Dekker; 2000. 971 p

[3] Kelley WG, Peterson AC. The Theory of Differential Equations: Classical and Qualitative. New York, NY, USA: Springer; 2010. 413 p. DOI: 10.1007/978-1-4419-5783-2

[4] Kelley WG, Peterson AC. Difference Equations, Second Edition: An Introduction with Applications. Burlington, MA, USA: Academic Press; 2001. 403 p

[5] Bohner M, Peterson A. Dynamic Equations on Time Scales: An Introduction with Applications. Boston: Birkhäuser; 2001. 358 p. DOI: 10.1007/978-1-4612-0201-1

[6] Bohner M, Peterson A. Advances in Dynamic Equations on Time Scales. Boston: Birkhäuser; 2003. 348 p. DOI: 10.1007/978-0-8176-8230-9

[7] Agarwal RP, O'Regan D, Saker SH. Oscillation and Stability of Delay Models in Biology. Switzerland: Springer; 2014. 340 p. DOI: 10.1007/978-3-319-06557-1

[8] Kyrychko YN, Hogan SJ. On the use of delay equations in engineering applications. Journal of Vibration and Control. 2010;**16**(7–8):943-960. DOI: 10.1177/1077546309341100

[9] Nehari Z. On a nonlinear differential equation rising in nuclear physics. Proceedings of the Royal Irish Academy. 1963;**62**:117-135

[10] Ruan S, Wolkowicz GSK, Wu J. Differential Equations with Applications to Biology. Providence, RI, USA: AMS; 1999. 509 p

[11] Wu L, Lam H, Zhao Y, Shu Z. Time-delay systems and their applications in engineering. Mathematical Problems in Engineering. 2015;**2015**:246351. 1-3 p

[12] Bartolini G, Pvdvnowski P. Approximate linearization of uncertain nonlinear systems by means of continuous control. In: Proceedings of the 30th Conference on Decision and Control; Brighton, England; 1991

[13] Bartolini G, Pisano A, Punta E, Usai E. A survey of applications of second-order sliding mode control to mechanical systems. International Journal of Control. 2010;**76**(9–10):875-892

[14] Bartolini G, Ferrara A, Usai E. Applications of a sub-optimal discontinuous control algorithm for uncertain second order systems. International Journal of Robust and Nonlinear Control. 1997;7:299-319

[15] Esmailzadeh E, Mehri B, Nakhaie J. Periodic solution of a second order, autonomous, nonlinear system. Nonlinear Dynamics. 1996;**10**:307-316

[16] Fonda A, Manásevich R, Zanolin F. Subharmonic solutions for some second-order differential equations with singularities. SIAM Journal on Mathematical Analysis. 1993;**24**(5):1294-1311

[17] Li X, Zhang Z. Periodic solutions for some second order differential equations with singularity. The Zeitschrift für Angewandte Mathematik und Physik. 2008;**59**(400):415

[18] Fowler RH. The form near infinity of real, continuous solutions of a certain differential equation of the second order. The Quarterly Journal of Mathematics. 1914;**45**:289-350

[19] Fowler RH. The solution of Emden's and similar differential equations. Monthly Notices of the Royal Astronomical Society. 1930;**91**:63-91

[20] Fowler RH. Some results on the form near infinity of real continuous solutions of a certain type of second order differential equations. Proceedings of the London Mathematical Society. 1914;**13**:341-371. DOI: 10.1112/plms/s2-13.1.341

[21] Fowler RH. Further studies of Emden's and similar differential equations. The Quarterly Journal of Mathematics. 1931;**2**:259-288

[22] Arthur AM, Robinson PD. Complementary variational principle for $\nabla^2 = f(\Phi)$ with applications to the Thomas–Fermaind Liouville equations. Mathematical Proceedings of the Cambridge Philosophical Society. 1969; **65**:535-542

[23] Davis HT. Introduction to Nonlinear Differential Integral Equations. Washington, D.C., USA: U.S. Atomic Energy Commission; 1960. (Reprint: New York, USA: Dover; 1962)

[24] Shevyelo VN. Problems methods and fundamental results in the theory of oscillation of solutions of nonlinear nonautonomous ordinary differential equations. In: 2nd All-Union Conference on Theoretical and Applied Mechanics Proceedings; Moscow; 1965. pp. 142-157

[25] Öztürk Ö. Classification schemes of nonoscillatory solutions for two-dimensional time scale systems. Mathematical Inequalities & Applications. 2017;**20**(2):377-387

[26] Öztürk Ö. On oscillatory behavior of two-dimensional time scale systems. Adv. Difference Equ. 2018;**2018**:18

[27] Öztürk Ö. On oscillation of two-dimensional time-scale systems with a forcing term. Turkish Journal of Mathematics. 2018;**42**:312-319

[28] Kac V, Cheung P. Quantum calculus. In: Universitext. New York: Springer-Verlag; 2002. 112 p. DOI: 10.1007/978-1-4613-0071-7

[29] Akın E, Öztürk Ö. On Nonoscillatory Solutions of Two Dimensional Nonlinear Dynamical Systems, Book Chapter of Dynamical Systems—Analytical and Computational Techniques. Rijeka, Croatia: IntechOpen; 2017

[30] Zeidler E. Nonlinear Functional Analysis and its Applications–I: Fixed Point Theorems. New York: Springer-Verlag; 1986. 909 p. DOI: 10.1007/978-1-4612-4838-5

[31] Knaster B. Un théorème sur les fonctions d'ensembles. Annales de la Societe Polonaise de Mathematique. 1928;**6**:133-134

[32] Li WT. Classification schemes for positive solutions of nonlinear differential systems. Mathematical and Computer Modelling. 2001;**36**:411-418

[33] Li WT. Classification schemes for nonoscillatory solutions of two-dimensional nonlinear difference systems. Computers and Mathematics with Applications. 2001;**42**:341-355

[34] Agarwal RP, Li WT, Pang PYH. Asymptotic behavior of nonlinear difference systems. Applied Mathematics and Computation. 2003; **140**:307-316

[35] Cheng S, Li HJ, Patula WT. Bounded and zero convergent solutions of second

order difference equations. Journal of Mathematical Analysis and Applications. 1989;**141**:463-483

[36] Cecchi M, Došlá Z, Marini M. Unbounded solutions of quasilinear difference equations. Computational and Applied Mathematics. 2003;**45** (6–9):1113-1123

[37] Anderson DR. Oscillation and nonoscillation criteria for two-dimensional time-scale systems of first order nonlinear dynamic equations. Electronic Journal of Differential Equations. 2009;**2009**(24):1-13

[38] Ciarlet PG. Linear and Nonlinear Functional Analysis with Applications. Philadelphia, PA, USA: SIAM; 2013. 832p

[39] Zhang X, Zhu S. Oscillation for a nonlinear dynamic system with a forced term on time scales. Abstract and Applied Analysis. 2014;**2014**:1-6

[40] Öztürk Ö, Akın E. On nonoscillatory solutions of two dimensional nonlinear delay dynamical systems. Opuscula Mathematica. 2016; **36**:651-669

[41] Öztürk Ö, Akın E. Nonoscillation criteria for two-dimensional time-scale systems. Nonautonomous Dynamical Systems. 2016;**3**:1-13

[42] Saker SH. Oscillation of second-order forced nonlinear dynamic equations on time scales. Electronic Journal of Qualitative Theory. 2005;**23**: 1-17

[43] Özgün SA, Zafer A, Kaymakçalan B. Gronwall and Bihari type inequalities on time scales. Advances in Difference Equations-Ny (Veszprém, 1995). 1997; **1997**:481-490

[44] Öztürk Ö, Akın E. Classification of nonoscillatory solutions of Emden–Fowler dynamic equations on time scales. Dynamic Systems and Applications. 2016;**25**:219-236

[45] Öztürk Ö, Akın E, Tiryaki IU. On nonoscillatory solutions of Emden-Fowler dynamic systems on time scales. Univerzitet u Nišu. 2017;**31**(6): 1529-1541

[46] Dŏslá Z, Vrkŏc I. On an extension of the Fubini theorem and its applications in ODEs. Nonlinear Analysis. 2004;**57**: 531-548

[47] Cecchi M, Dŏslá Z, Marini M. On oscillation and nonoscillation properties of Emden-Fowler difference equations. Central European Journal of Mathematics. 2009;**7**(2):322-334

Section 2

From the Atomic Scale to the Vibration of a Building

Chapter 4

Time-Domain Simulation of Microstrip-Connected Solid-State Oscillators for Close-Range Noise Radar Applications

Vladimir Yurchenko and Lidiya Yurchenko

Abstract

We develop time-domain approach for simulation of microstrip-connected extremely-high frequency (EHF) solid-state oscillators for close-range radars, including ultrashort-pulse, ultrawide-band (UWB), and noise radars. The circuits utilize high-speed GaN-based active devices such as Gunn diodes (GD) and resonant-tunneling diodes (RTD) capable of operating with enhanced power output. Microstrip interconnects produce time-delay coupling in the system that can create a complicated nonlinear dynamics of oscillations. The circuits can generate self-emerging trains of ultra-short EHF pulses emitted into an open microstrip section for further radiation. The arrays of active devices connected in either parallel (star-case) or series (ladder-case) type of circuits were simulated. Options for generation of chaotic signals in this kind of systems have been considered. An infrared-microwave (IR-EHF) oscillator linked to the resonant antenna was simulated. The oscillator consists of an RTD-driven laser diode (LD) joint to the EHF resonant antenna with a short piece of microstrip section. The oscillator can generate both the EHF pulse radiation and the EHF modulated IR pulses. Both kinds of radiation can be emitted in the free space as the trains of correlated IR-EHF radar pulses. Arrays of oscillators can be used for enhancing the power output of the system.

Keywords: time-domain simulations, solid-state oscillator, THz, millimeter wave, time-delay, chaos, distributed systems, active devices, Gunn diode

1. Introduction

Emerging demands for the EHF oscillators capable of generation of ultra-short pulses and complicated waveforms including chaotic oscillations lead to the development of new approaches to the design and analysis of oscillator systems. The EHF oscillators are of interest for numerous applications. Significant part in these applications belongs to radar systems including, particularly, close-range and noise radars, which require ultrashort-pulse, ultrawide-band, and noise oscillation sources [1, 2]. There are different kinds of the EHF oscillators ranging from microwave power tubes (klystrons, gyrotrons, backward-wave tubes, etc.) to solid-state devices (transistor-based circuits, Gunn diodes, etc.) [3, 4], of which only the latter are discussed in this chapter.

Design of oscillators and circuits is conventionally made in frequency domain. A significant contribution to the design was made by Kurokawa [5, 6] through advancing the negative resistance oscillator concepts and developing stability analysis methods. He developed the impedance approach to the analysis of oscillator systems that makes it possible to design, in particular, multi-device circuits with spatial power combining [6].

Numerous advances to the design and analysis of oscillators have been made in the following years [7–9]. Significant developments are the extension of the frequency-domain analysis for the account of nonlinear characteristics of active devices [7], the analysis of different impedance and admittance formulations in the oscillator design [8], the application of hybrid harmonic-balance approach [9], etc. A vast literature exists on the design of oscillators with frequency-domain methods.

In this line, time-domain oscillator analysis is not a common practice. In order to deliver essential information about the oscillators and their dynamics, time-domain analysis requires huge amount of numerical simulations of complicated oscillatory systems, which have to be made in a broad range of oscillator parameters.

Despite this difficulty, there are circumstances when such an analysis is a necessity, since no alternative approach can provide adequate information on the oscillator dynamics in the relevant cases. These are the cases when ultrashort-pulse, ultrawide-band, and noise oscillation signals have to be generated [1, 2]. The problem exacerbates when signals should have extremely broad frequency spectrum extended in the EHF and THz bands.

Design of generation and transmission systems for this kind of signals inherently requires the time-domain approach [10]. For passive components like antennas, valuable contributions to mathematics of time-domain modeling that concerns ultrashort-pulse and ultrawide-band signals have been made [11, 12]. The oscillators are, however, much too complicated unstable and nonlinear systems for the efficient simulations. Nonetheless, time-domain modeling is, in fact, the most meaningful approach to the design of oscillators generating this kind of signals [13, 14], though the frequency-domain methods can also be helpful [15].

A practical way of making progress in the analysis of these oscillators is to consider simplified models, which, despite their simplicity, represent essential features of real systems. An important feature of oscillators in the EHF and THz bands is their distributed character. Even though active devices and other discrete elements may be small, their assembly into an operating circuit with extended interconnects, resonators, and antenna components makes the entire system to be comparable to the radiation wavelength.

Thus, time delay arises, essentially, due to the delayed coupling between the components that makes the circuit to operate as a distributed system. Time delay leads to complicated dynamics and, often, to the dynamical chaos in nonlinear systems [16] that makes time-delay oscillators to be attractive devices for numerous applications.

A particularly useful simplification arises when making clear distinction between discrete and distributed components and defining the model where discrete units (circuit elements or blocks of elements) are joined by transmission lines (waveguides, microstrips, etc.) in a way that qualitatively represents the actual connectivity of components in the entire system. Then, discrete blocks can be simulated by local equations in time domain and the effects of transmission lines can be accounted by readily available analytic solutions of simplified wave equations. The entire system is then suitable for reasonably efficient time-domain simulations.

We applied the approach to the time-domain analysis of a range of different distributed circuits with active devices where we assumed that the transmission lines are, typically, the microstrip sections and the active devices are either the Gunn diodes or the resonant tunneling diodes in that or another circuit [17–25]. Transmission lines introduce time delays in the coupling between discrete units that makes the entire circuits to operate as the time-delay oscillators.

Our microstrip-based oscillator models are qualitatively different from other time-delay oscillators usually considered [4]. The difference is that, instead of using pre-defined phase delays in the feedback circuits, we consider time-delays that emerge self-consistently as a result of backward and forward EM wave propagation along the transmission lines with account of their scattering and interference with other process, making the effects particularly complicated. The systems take into account the fact that, at the frequencies of the EHF and THz range, i.e., for the millimeter and sub-millimeter waves, time-delays become unavoidable due to the extended structure of oscillator circuits.

The following Sections present overview of basic results obtained in our time-domain simulations of extended transmission line time-delay oscillators.

2. Kinds of circuits and forms of oscillations

We consider solid-state oscillators that can be presented as a combination of both the lumped units (lumped circuit blocks) and distributed microstrip sections (pieces of transmission lines) of different configurations. Microstrip sections provide interconnects between the lumped units and produce time-delay in the coupling between different circuit components.

The lumped units are built up of discrete active and passive devices whose interconnects within each block are of infinitesimal length as compared to the typical wavelength λ_0 of the electromagnetic (EM) waves emerging in the system as a result of complicated self-oscillation process. For this reason, there is no time delay arising due to signal propagation between discrete elements, including active devices, within each lumped unit. There are neither special time-delay devices of other kinds included in the lumped units. The units being used are, in fact, rather simple pieces of circuits made up of active devices (Gunn diodes, avalanche diodes, resonant tunneling diodes) and passive elements (resistors, capacitors, and inductances).

The distributed sections are the pieces of microwave transmission lines (e.g., microstrip lines as representative elements or any other waveguide structures). Transmission lines (TL) provide time-delay coupling between the lumped units. Time delay appears in the coupling between the lumped units connected by any TL section because of some time needed for the EM wave propagation along the section between the units. The time delay has to be accounted in the analysis when the length d_n of the relevant TL section identified by index n is not too small as compared to the typical wavelength λ_0 of the EM waves propagating along the section.

Schematics of a few circuits being considered are presented in Sections 5, 7, and 8 below. Depending on the kind of circuits, different types of nonlinear oscillation can be excited in time-delay systems. The key elements in these distributed systems are the active blocks that contain one or another kind of solid-state active devices.

We consider the EHF solid-state devices such as Gunn diodes or the resonant tunneling diodes (RTD) that can operate in a broad range of frequencies varying from, essentially, 10GHz to about 1 THz and more. The operation of these devices and oscillators is best understood in terms of the negative resistance oscillator concepts [7, 8]. Typically, we consider the Gunn diode circuits in our models, though one example of RTD system is discussed in the Section 8. In practice, the most common are the GaAs Gunn diodes but GaN devices are now of greater interest due to their potential for high-power and high-frequency operation.

The Gunn diodes are simulated using the approximation of limited space-charge accumulation (LSA) mode. In this mode, the strong-field domain in the Gunn diode is bounded to the surface electric contact and can only oscillate near the contact rather than travel through the entire structure. Then, the oscillation frequency of the Gunn diode can vary in a broad range and achieve rather high values.

Using the LSA approximation, the device operation can be described in terms of the given current-voltage characteristics with negative differential resistance (NDR) region. In this model, the current-voltage characteristics of typical Gunn diodes, e.g., GaAs diodes, can be presented in the following form [17–24].

$$G(e) = G_0 F(e) \tag{1}$$

where $G(e)$ is the diode current in relative units, $F(e)$ is the function defining the shape of the current-voltage characteristics, $G(e) = G_0 I(e)/I_0$, $G_0 = Z_0 I_0/V_0$, $e = V/V_0$, I_0 and V_0 are the scaling factors for the diode current $I(e)$ and voltage V, respectively, and Z_0 is the microstrip wave impedance (I_0 and V_0 parameters are specified by the Gunn diode threshold current and voltage, respectively).

The Gunn diode self-excitation begins when the voltage V falls in the NDR region. We accept the intrinsic impedance of microstrip lines to be $Z_0 = 50$ Ohm and use $G_0 = 2$ as a typical Gunn diode parameter (typically, $I_0 = 0.2$A, $V_0 = 5$ V for GaAs and $I_0 = 1.2$ A, $V_0 = 30$ V for GaN structures).

The formulation in terms of the current-voltage characteristics, which is typical for the LSA approximation, means that the diode is capable of instant response to any external signal. The operation of such a diode is, formally, not limited from above by any high frequency value. In reality, though, the high-frequency operation is limited by the diode intrinsic capacitance C and the inductance L of the mounting contacts. These parameters define the natural intrinsic frequency of the Gunn diode when it is mounted in one or another way in the transmission line. Typically, due to quite noticeable value of inductance, this intrinsic frequency is lower than the highest frequency accessible for the diode operation.

In our models, we consider both kinds of approximations when the Gunn diode is either not limited in the oscillation frequency or, on the contrary, is characterized by intrinsic capacitance and inductance, which impose the limit on the diode operation frequency. In the latter case, the intrinsic capacitance and inductance are defined as the effective components directly connected to the Gunn diode within the lumped active unit.

Early models, for simplicity, did not account for the diode capacitance and inductance, thus, ignoring the diode frequency limit. The approximation allowed us to significantly simplify the original problems and reduce them to the forms which are more accessible for numerical simulations. In this way, we could consider self-excitation and nonlinear dynamics of the EM field oscillations in a closed two-dimensional (2D) rectangular cavity with an active wall [17, 18] and self-emergence of trains of pulses emitted from one-dimensional (1D) cavity with such a wall on the one side and a dielectric plate as a semi-transparent mirror on the other side that makes the cavity an open resonator [19]. These simulations are discussed in Section 4.

Later models took into account the intrinsic frequencies of active devices. When applied to 1D microstrip circuits and the networks of circuits with either parallel or series connections of branches, we could observe and analyze a series of new effects in self-oscillations of these circuits [20–25]. The circuits and relevant effects are considered in Sections 5–8.

The dominant effect in these circuits, apart from conventional continuous wave (CW) generation, is the excitation of trains of short radio-frequency (RF) pulses. When using high-speed devices such as GaAs and GaN Gunn diodes in circuits with proper other parameters such as a high resonator frequency and a short length of microstrip sections, one can achieve self-excitation of short trains of ultra-short EHF pulses, which are emitted into an infinite section of another transmission line. Self-developing transitions between either the CW or pulse modes of EHF oscillations are possible, which depend on both the operation conditions and prehistory of oscillations, thus, revealing the hysteresis and bistability effects in the time-delay EHF oscillators being considered.

3. Mathematical models and simulation techniques

Time-domain analysis of nonlinear oscillator systems is based on computer simulations for the numerical solutions of oscillatory equations. Typically, ordinary differential equations (ODE) for lumped systems or partial differential equations (PDE) for distributed structures are in use. For more complicated cases, integral-differential, difference-differential, difference-integral-differential, and other kinds of equations may be needed for the adequate modeling of real oscillator systems. The term difference-differential means the equations that account for the finite delay in their arguments that appear in some terms of the equations. The equations of this kind are also called the equations with deviating arguments.

In our models, the delay in the electromagnetic coupling between spatially separated blocks creates multiple time delays in the evolutionary equations that can be derived for the entire oscillator system. The equations are formulated for the electromagnetic field, current, and voltage quantities as functions of time and spatial coordinates in the given circuits. Using various transformations, the equations are reduced to some other forms mentioned above to make their numerical solution more accessible.

Specific examples of equations obtained in this way for some of the models are presented in the following Sections. Typically, the equations are reduced to one or another version of a set of ODEs with time-delay arguments in their terms that describe the EM wave coupling in the distributed microstrip circuits.

For the accurate solution of these equations, we applied a highly efficient and reliable Dormand-Prince method of the (5,3) order of accuracy [26]. A publicly available software code of the method was amended with our extension that provided the dynamical storage of dense output of a solution in a long period of time in the past so as all the time-delayed values were available.

Since the past values are requested by solver at some unknown time nodes in the storage domain that may exceed the formal time-delay value, we created sufficiently large storage domain, applied polynomial interpolation of a high and controllable accuracy that uses a big domain and many nodes around the requested node, and provided a special control that no limitations are broken.

Thus, we obtained a unique, highly accurate and reliable software tool for time domain simulations of complicated nonlinear dynamics of time-delay oscillators that may exhibit dynamical chaos and other unstable transient, oscillatory, and evolutionary developments. The tool appeared to be highly efficient for solving the time-domain problems arising for our models of time-delay oscillatory systems.

More generic software tools could also be used for time-domain simulations of oscillator circuits. At present, SmartSpice simulation software is available for the engineering applications of time-domain circuit analysis [27] though applicability of generic techniques for time-delay problems and complicated circuits is limited.

The other circuit models that assumed an instant response of active devices (excluding the case of 2D cavity) lead to the difference-delay equations rather than time-delay ODEs. These equations are solved by direct iteration process.

In all the problems considered, the initial conditions were spatially uniform steady-state solutions that exist at the given initial parameters including the bias voltage V_B regardless whether the solution is stable or unstable. In case of a stable initial condition, when the device voltage V_G is out of the NDR region, the bias V_B is altered with a certain rise time so as to set the device voltage V_G in the NDR domain where the oscillations begin. In the other cases, when the initial condition is unstable due to the initial device voltage V_G is set in the NDR region, a minor fluctuation of the bias voltage is introduced that initiates the circuit self-excitation process.

4. Early models, dynamical chaos, and pulses

The early models assumed an instant response of active devices. In this way, they ignored any possible frequency limitations imposed by the limited operation speed of active devices characterized by the relevant intrinsic frequency f_G. So, any specific frequencies of emerging oscillations were defined by the passive circuit components and nonlinear character of the entire oscillator system.

The first model [17, 18] is presented by the 2D rectangular cavity ($0 < x < D$, $0 < z < A$) with perfectly conducting walls, of which one wall ($x = D$) is covered with an active semiconductor layer specified by nonlinear current-voltage ($I - V$) characteristics with NDR domain as defined in Eq. (1). The active layer is used as an approximation for, e.g., a dense array of active devices such as the Gunn diodes placed on one of the cavity walls inside the cavity. The quantities of interest are the electric field in the cavity $E_z(x, z, t)$ and the average field $U(t)$ at the active layer when the external voltage is applied to active devices and the electric field is set in the NDR region. Self-excitation is developing in the system when a minor fluctuation of the EM field is introduced in the cavity.

The evolution of self-excitation may occur in a different manner depending on the system parameters. An essential parameter is the coupling coefficient G_0 representing the strength of coupling of the electric field at the cavity wall $x = D$ with active devices and, implicitly, the maximum electric current in the active layer. In this way, the coefficient defines the EHF power that can be generated by active devices.

The main result obtained in this model is the emergence of the dynamical chaos of the EM field in the cavity when the coupling coefficient G_0 is made sufficiently large [17, 18]. To identify the chaos, three basic criteria have been used: (a) specific structure of the Poincare sections, (b) broad-band character of the power spectrum of oscillations and (c) sensitive dependence of solutions on the initial conditions. In addition, the wavelet analysis was applied to study the evolution of the excited field. The Morlet wavelets have been used to scale the frequency band $0.4 \leq fD/c \leq 6.4$ in the time interval of $375 \leq ct/D \leq 400$ where c is the speed of light.

Most of simulations were made with a square cavity ($A = D$). The square cavity often generated chaotic dynamics of the $E_z(x, z, t)$ field whereas a shorter cavity (e.g., $A = 0.8D$) showed a greater stability. Yet, the main factor controlling the emergence of chaos is the coupling coefficient G_0. When the coefficient is small, e.g., $G_0 = 0.1$ in case of the square cavity, only regular and, essentially, single-frequency oscillations appear. At the intermediate coupling, e.g., $G_0 = 1$, the co-existence of multi-frequency and chaotic generation is observed. Finally, at the very strong coupling when, e.g., $G_0 = 30$, chaotic oscillations dominate.

Simulations in this 2D model were made using the series expansion of the cavity field in spatial modes with time-dependent expansion coefficients in a way, which is conceptually similar to formulations developed in [11]. In this way, a set of nonlinear ODEs was obtained for the expansion coefficients as unknown functions. The active layer $I - V$ characteristics entered this formulation through the boundary conditions imposed at the active wall $x = D$, where they played a role of nonlinear impedance boundary conditions formulated in time domain. Other approaches using, e.g., Green's function formulations and taking into account non-instant response of active devices, should also be explored.

The second model [19] is formulated as a 1D open-cavity problem where the 1D cavity $(0 < x < D)$ is formed by the first wall at $x = D$, which is covered with the same instant-response active layer as explained above, and the second wall at $x = 0$, which is made as a thin dielectric layer of thickness d $(-d < x < 0)$ that operates as a resonant semi-transparent mirror for the cavity field.

In this case, the problem formulated for the electric field $E_z(x, t)$ that exists in both the cavity, the dielectric mirror, and the half-space outside the cavity, is reduced to a single, though, complicated, delay-difference equation with multiple time delays, which is formulated for the auxiliary function $g(\tau)$ where $\tau = ct/D$ is the time variable t in the relative units. The other functions of interest such as the oscillation waveform $U_1(\tau)$ emitted through the dielectric mirror are also defined via the function $g(\tau)$ [19].

The main result obtained for these structures is that, typically, a train of the EHF pulses is self-excited in the 1D oscillator. The basic frequency of the EHF oscillations in these pulses is defined by the thickness d and refractive index n of dielectric mirror. No intrinsic frequency of active devices is present in this problem since active devices are specified by an instant response and impose no frequency limits on the emerging self-oscillations. In other cases, depending on the parameters, trains of baseband pulses with no carrier frequency are self-excited.

Characteristic frequencies of the EHF pulses emerging in these structures would be, e.g., $f = 64$ GHz ($\lambda_0 = 4.7$ mm) at the pulse duration $t_P = 0.17$ ns (the pulse frequency width $f_P = 1/t_P = 6$ GHz) and the pulse repetition frequency $f_{REP} = 3$ GHz should the devices were capable of operation at these frequencies, the dielectric mirror made of a MgF_2 wafer of thickness $d = 1$ mm having the refractive index $n = 2.345$ at the extremely low loss tangent $\tan(\delta) = 5 \cdot 10^{-5}$ [28], and the resonator length chosen to be $D = 25$ mm.

5. Active circuit with a remote resonator antenna

The model that removes an essential limitation of instant response of active devices was investigated in detail in [20]. The model is rigorously formulated in terms of the 1D microstrip distributed circuit with spatially separated active and passive components, **Figures 1** and **2**.

The circuit consists of four basic parts which are (i) an active block G_1 made as a lumped unit with Gunn diode as an active device, (ii) a remote resonator A_0 (we assume this is an LC circuit implemented as a lumped block), (iii) a section of microstrip line of length D connecting the active block and the resonator (this is a distributed part of the circuit), and (iv) an infinite section of microstrip line connected to the resonator. The latter allows the waves excited in the circuit to be radiated towards the infinity, thus, simulating the radiation into the free space, with resonator block A_0 operating also as an antenna unit.

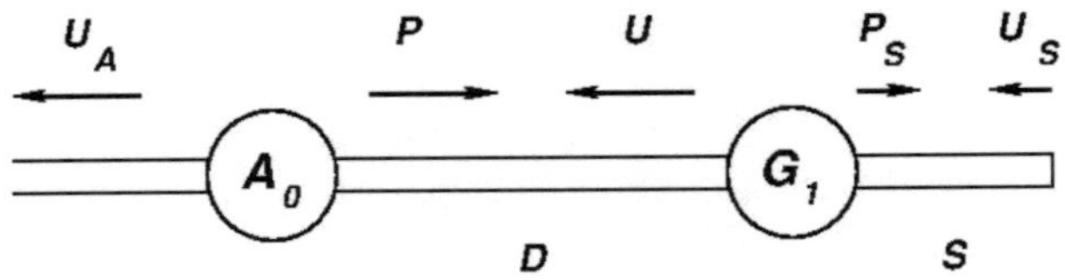

Figure 1.
Schematics of a microstrip circuit with Gunn diode active block G_1 and a resonator antenna block A_0.

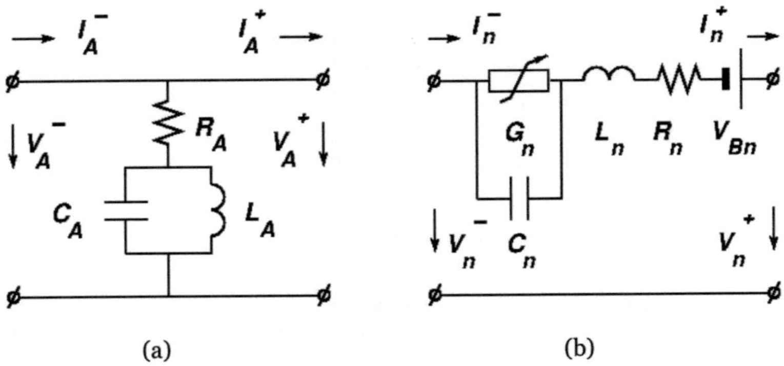

Figure 2.
Schematics of (a) passive and (b) active blocks of **Figure 1** *($n = 1$).*

Figures 1 and **2** show a more generic version of the system considered in [20]. The system of a generic kind includes the stub of the length S that facilitates the emergence of self-oscillations and the lumped circuit resistors R_A and R_n ($n = 1$) that simulate absorption losses (equations below and the results are presented for a simpler case of $S = 0$ and $R_A = R_1 = 0$).

The governing differential equations and boundary conditions are obtained by applying the wave equations to the transmission line sections and the Kirchhoff circuit equations to the diode and resonator blocks. The wave equations describe the voltage waves P, U, and U_A that propagate to the right and to the left in the microstrip sections as shown in **Figure 1**, respectively. They are the unknown functions to be found.

The Kirchhoff equations define boundary conditions imposed at the contact points of both the Gunn diode and resonator circuits. They are formulated in terms of the voltage and current values at the contact points $e^{\pm}$, $e_A^{\pm}$, $i^{\pm}$, $i_A^{\pm}$, which are defined via the unknown waveforms P, U, and U_A. The radiation boundary condition is applied that ensures no incoming waves in the open microstrip section (microstrips are assumed lossless and free of dispersion).

The initial condition is imposed as the state of no oscillations when the diode voltage is set outside the NDR region by the source voltage e_{B_0}. Self-excitation appears when the Gunn diode is driven into the NDR region by increasing (decreasing) the source voltage $e_B(\tau)$ with a certain rise (fall) time T_R (T_F).

The time and space variables are used in relative units $\tau = ct/a$ and $x = \tilde{x}/a$ where t and $\tilde{x}$ are the original time and space variables, respectively, c is the speed of wave in the microstrip line, $d = D/a$ is the length of microstrip section in relative units, and a is the spatial scale used for normalization.

The equations for this model are reduced to a set of ODEs with time delays

$$U''(\tau) + U''(\tau - 2d) - U''_A(\tau - d) + \omega_L\left[U'(\tau) - U'(\tau - 2d) + U'_A(\tau - d) + e'_B(\tau)\right]$$
$$+ \omega_G^2\left[U(\tau) + U(\tau - 2d) - U_A(\tau - d) - G(e(\tau)) + G(e_0)\right] = 0, \tag{2}$$

$$U_A''(\tau) + 2\omega_{C_A}\left[U_A'(\tau) - U'(\tau - d)\right] + \omega_A^2 U_A(\tau) = 0, \tag{3}$$

$$\begin{aligned} e(\tau) &= e_B(\tau) - U(\tau) + U(\tau - 2d) - U_A(\tau - d) \\ &- \omega_L^{-1}[U'(\tau) + U'(\tau - 2d)] - U_A'(\tau - d)] \end{aligned} \tag{4}$$

where $U(\tau)$ and $U_A(\tau)$ are the unknown wave functions, $e(\tau)$is the Gunn diode voltage at the time τ, $\omega_L = aZ_0/cL$, $\omega_{C_A} = a/cZ_0C_A$, $e_0 = e_{B_0}$, $\omega_G^2 = (a/c)^2/LC$, $\omega_A^2 = (a/c)^2/L_AC_A$, and $G(e(\tau))$ is the Gunn diode current defined by Eq. (1). Here, index A denotes the values related to the resonator antenna block, index $n = 1$ is dropped, and the other related angular frequencies are $\omega_{L_A} = {\omega_A}^2/\omega_{C_A}$ and $\omega_C = {\omega_G}^2/\omega_L$, respectively.

6. Bistability, hysteresis, and trains of the EHF pulses

Simulations of oscillators with active circuits and remote resonator antennas revealed the existence of two oscillation modes, of which one mode is the CW oscillations and the other is the trains of the EHF pulses. The emergence of one or another oscillation mode depends on the bias voltage of the Gunn diode circuit. The EHF pulse mode arises when the Gunn diode operating voltage V_{GO} exceeds the lower bound V_1 of the NDR region but the excess is not significant so as the voltage V_{GO} does not fall deep in the NDR region. The CW oscillations, on the contrary, appear when the operating voltage V_{GO} is set deep in the NDR region [20].

An important feature of the effect is the co-existence of both the pulse and the CW modes of oscillations in some range of operating voltages. The oscillator can generate either the EHF pulses or CW oscillations at the same operating voltage V_{GO} when the latter falls in an intermediate domain $V_{B1} < V_{GO} < V_{B2}$ (PC-domain) inside the NDR region $V_1 < V_{GO} < V_2$. At smaller values of operating voltage that fall out of this domain but inside the NDR region, $V_1 < V_{GO} < V_{B1}$, the trains of the EHF pulses are self-excited (P-domain). At greater values, $V_{B2} < V_{GO} < V_2$, the CW oscillations are generated (C-domain).

The effect means bistability of the oscillation modes in the PC-domain. The kind of the mode being excited depends on the history of bias variations, i.e., on the way the oscillator is driven to the operating voltage. In this process, the oscillator reveals an hysteresis in switching between different modes. When the oscillator is driven into the PC-domain through the P-domain starting from small values, the EHF pulses are self-excited. On the contrary, when the oscillator is driven through the C-domain starting from large values, the CW oscillations are generated.

The PC-domain borders also depend on the direction of driving the operating voltage into this domain. So, there are, in fact, two kinds of P, C, and PC domains, which could be labeled, e.g., as P-up and P-down, C-up and C-down, and PC-up and PC-down domains. The effect is illustrated in **Figure 3** that shows the I-V curve of the Gunn diode in relative units $G(e)$ and the relevant domains of different oscillation modes.

Depending on the speed of driving the diode into the relevant stable domain of either the P or C kind, one can excite initially the oscillation mode, which is not intrinsic for that domain, e.g., CW oscillations in the P-domain. This mode is, however, unstable and after a certain period of time it gradually turns into the mode, which is intrinsic for the given domain, e.g., into the pulse mode in P-domain. Similarly, pulses, which could initially be excited in C-domain, gradually transform into the CW oscillations, which remain stable in this domain. The latter process is illustrated in **Figure 4**.

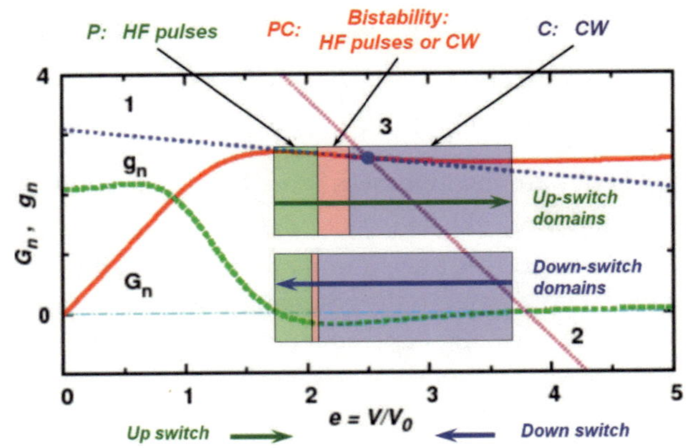

Figure 3.
Gunn diode current-voltage characteristics $G_n(e)(n = 1)$, differential conductance $g_n(e) = dG_n(e)/de$, two options for the load lines (curves 1 and 2), the operation point 3, and the voltage regions P, C, and PC, that correspond to the emergence of different oscillation modes in the NDR region $e_1 < e < e_2$ ([20]; licensed under a creative commons attribution (CC BY) license).

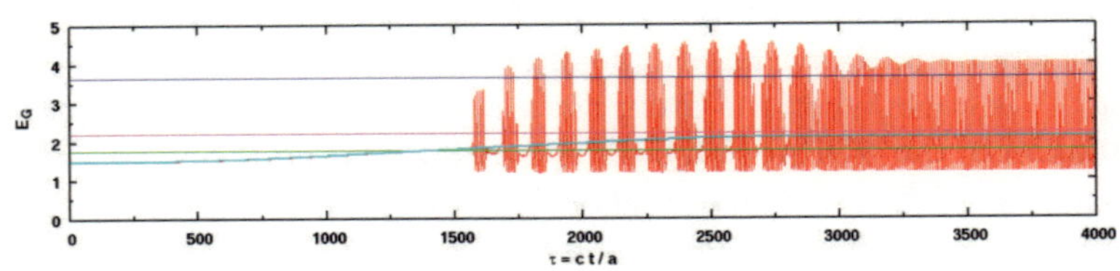

Figure 4.
Transition from the initial pulses to CW oscillations in the circuit with $d = 20$ at a very slow switching the bias voltage up to $e_B = 2.1$ in the C-up domain ([20]; licensed under a creative commons attribution (CC BY) license).

Let us now consider the properties and the mechanism of emergence of the EHF pulses. The time length of each pulse t_P appears to be equal to the interval of time between pulses Δt_P and each of them is equal to the time of the return trip of the EM signal between the active block G_1 to the resonator A_0. Thus, the EM pulse length $L_P = ct_P$ is twice the length of microstrip section connecting the Gunn diode and the resonator, $L_P = 2D$ at $t_P = \Delta t_P = 2D/c$.

The pulse carrier frequency ω is defined by the intrinsic frequencies of both the active block ω_G and the resonator ω_A. The condition for the emergence of a clear train of pulses is the coincidence of frequencies ω_G and ω_A. The length D of microstrip section has to be sufficiently large for the pulse duration t_P to be much greater than the oscillation period $T = 2\pi/\omega$. An example of a perfect train of the EHF pulses is shown in **Figure 5**.

In relative units, our simulations were made, typically, at the parameter values $\omega_A = \omega_G = 1$, $\omega_{C_A} = \omega_C = 10$, $\omega_{L_A} = \omega_L = 0.1, G_0 = 2$, and the microstrip length d chosen in the range of $d = 10 - 200$. The emerging radiation wavelength in relative units was $\lambda \approx 8$. When the normalization length a is chosen to be $a = 1$ mm, this corresponds to the oscillation frequency of $f \approx 37.5$ GHz. Then, $\omega_C = 10$ and $\omega_L = 0.1$ at $Z_0 = 50$ Ohm correspond to the capacitance $C = 0.07$ pF and the inductance $L = 0.17$ nH.

The emergence of one or another oscillation mode depends on the length d of the microstrip section between the Gunn diode and the remote resonator. The longer is the section, the easier trains of pulses are excited. Since the oscillations arise and decay at a very short time, the emergence of short pulses is possible at sufficiently short values of d, e.g., $d = 10$ at the wavelength $\lambda = 8$.

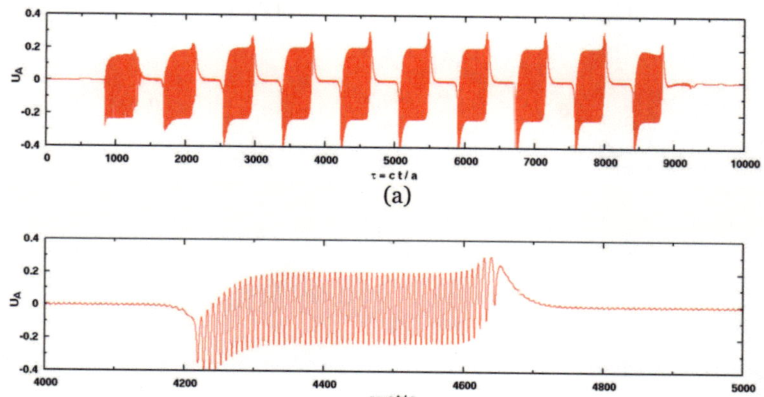

Figure 5.
(a) Train of the EHF pulses and (b) the shape of a single pulse generated by the circuit with $d = 200$ at the oscillation period $T \approx 8$ when switching the bias voltage up to $e_B = 2.0$ at the rise-time $T_R \approx 400$ ([20]; licensed under a creative commons attribution (CC BY) license).

The formation of trains of the EHF pulses can be explained as follows. If the circuit is designed so that excitation is possible with no resonator at the site A_0, the oscillations appear and continue for the time t_P until the feedback signal returns from the resonator to the active block.

Then, if the conditions are so that oscillations cannot exist with both the active block and the remote resonator engaged, the oscillations cease for the period of time Δt_P when the active block receives a feedback signal and, therefore, "feels" the presence of the resonator. After that period of time, the feedback disappears, the active block does not "feel" any resonator again, and a new pulse of oscillations arises.

If the system is so that oscillations can exist in the presence of resonator, the oscillations, once appeared, would continue as a steady-state process. In this case, CW oscillations are excited whose frequency is defined by the Gunn diode and the remote resonator circuits.

The effects described above are tightly connected to a general problem of oscillation quenching and collective behavior of oscillators [29]. They are also related to bifurcations observed in square-wave switching in delay-coupled semiconductor lasers [30]. It is clear that, despite essentially different governing equations, the common feature of time-delay coupling leads to similar consequences in terms of nonlinear dynamics of microwave oscillations and optical polarization in these cases.

The generation of trains of pulses discussed above is obtained in the model that accounts for non-instant response of devices caused by the capacitance and inductance of active units [20]. These simulations confirm the conclusions of a simpler model [19] that the excitation of trains of the EHF pulses is a generic property of those oscillators, which are specified by remote location of their resonator structures.

7. Parallel and series connections of active circuits

One can imagine many kinds of active circuits and distributed networks of microstrip connected discrete circuits. There are two basic types of connection of active circuits, which are the parallel and series connections, **Figures 6** and 7, respectively.

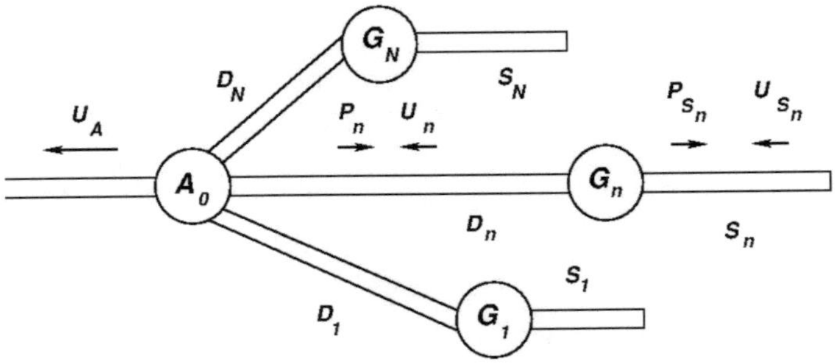

Figure 6.
Schematics of a parallel connection of active circuits where, in distinction from [21], different lengths of all microstrip sections D_n and stubs S_n are assumed as needed for quasi-chaotic oscillations.

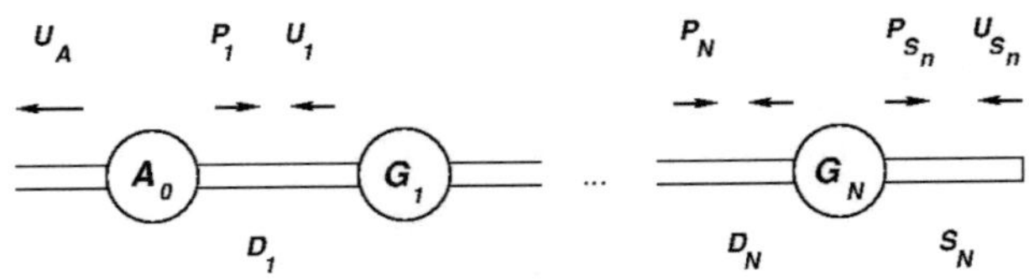

Figure 7.
Schematics of a series connection of active circuits where, unlike the case in [22], different lengths of different sections D_n are considered.

The parallel connection [21] is conceptually similar to the case of microwave cavity coupling of devices with spatial power combining considered by Kurokawa [6]. Spatial power combining is an important issue in this topic [31]. Our time-domain simulations of microstrip circuits connected in parallel confirm a possibility of increasing the total power output of the system proportional to the number of circuits N until a certain limiting value $N_{\max}$. At the same time, as typical for the circuits with a remote resonator A_0 connected to the active blocks G_n, trains of the EHF pulses can also be generated in these circuits.

The series connection of active circuits was also considered and the effect of nonlinear power combining was demonstrated [22]. The series connection of circuits appears to be less promising then the parallel connection since, due to the self-consistent evolution of the entire system, the basic oscillation frequency, typically, decreases with increasing the number of devices. The active blocks in [22] were different from those in [21] that, partially, could explain the effect. Nonetheless, the increase of the total length of the system in series connection is supposed to be the main reason of reducing the basic oscillation frequency.

Turning back to the parallel connection of active circuits and keeping in mind the explanation of the effect of pulsing presented in Section 6, we can consider a network of N parallel time-delay branches of Gunn diode circuits with different lengths of microstrip sections D_n [23]. With account of different times of arriving of time-delayed feedbacks in different branches and strong nonlinear mixing of oscillations in active devices, we can expect the development of complicated and, potentially, chaotic or quasi-chaotic oscillations.

As a test of this possibility, we considered a system of two parallel branches of identical active circuits presented in **Figure 2**, though of different and, generally, non-commensurable length of time-delay microstrip sections [23]. In this case, despite an apparent simplicity of the system, we observed complicated and, in the lower frequency bands, virtually quasi-chaotic nonlinear oscillations, **Figure 8**.

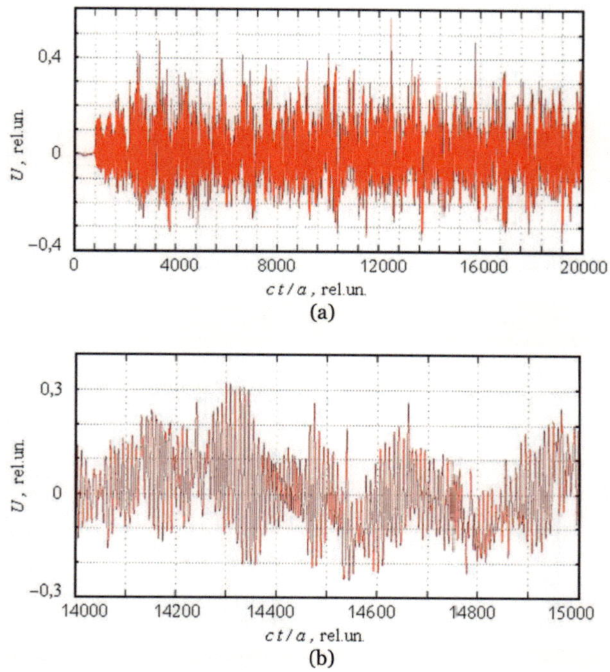

Figure 8.
(a) A quasi-chaotic signal radiated from a system of two Gunn diode circuits connected in parallel to the resonant antenna node by microstrip sections of the length $d_1 = 200$ *and* $d_2 = 266.67$*, respectively, when the basic radiation wavelength is* $\lambda = 8.6$ *and (b) the close-up view of a part of this signal [23].*

The effect is similar to the excitation of chaos in the 2D cavity [17, 18] or in the network of dispersive transmission lines where different frequency components take different times for the return of the feedback signal.

Quasi-chaotic character of the radiated wave in the low-frequency band was observed in both the Poincare sections and the auto-correlation functions of the emerging self-oscillations. When comparing auto-correlation functions of trains of the EHF pulses and quasi-chaotic signals arising under the relevant conditions, one can see a revival of correlations over the period of pulse repetition in the train of pulses and, on the contrary, a significant loss of correlation in the quasi-chaotic signal at all the times exceeding the basic period of oscillations, **Figure 9**.

Poincare sections plotted for the variables $U_A(\tau)$ and $dU_A(\tau)/d\tau$ show the presence of periodicity in the train of the EHF pulses and the lack of long-term periodicity in quasi-chaotic signals, **Figure 10(a)** and **(b)**, respectively.

The frequency spectrum of the quasi-chaotic signal shows the presence of chaotic components around the basic oscillation frequency and in the low-frequency band, **Figure 11**.

When the trains of the EHF pulses are excited, one can obtain rather short length of pulses. Since the process of turning on and off the EHF oscillation pulses is very short, one can obtained pulses that consist of just a few oscillations within each pulse when the microstrip sections between active circuits and resonant structures are sufficiently small, still being of nonzero length.

As an example, in a single-branch circuit of the kind shown in **Figures 1** and **2**, in case of $\omega_A = \omega_G = 1$, $\omega_{C_A} = 10$, $\omega_L = 0.1$ ($\omega_{L_A} = {\omega_A}^2/\omega_{C_A} = 0.1$, $\omega_C = {\omega_G}^2/\omega_L = 10$), $S = 0$, $R_A = R = 0$, and $G_0 = 2$, only two and four oscillations within each pulse were excited when the length of the microstrip section was $d = 10$ and $d = 20$, respectively, and the radiation wavelength was $\lambda \approx 8$.

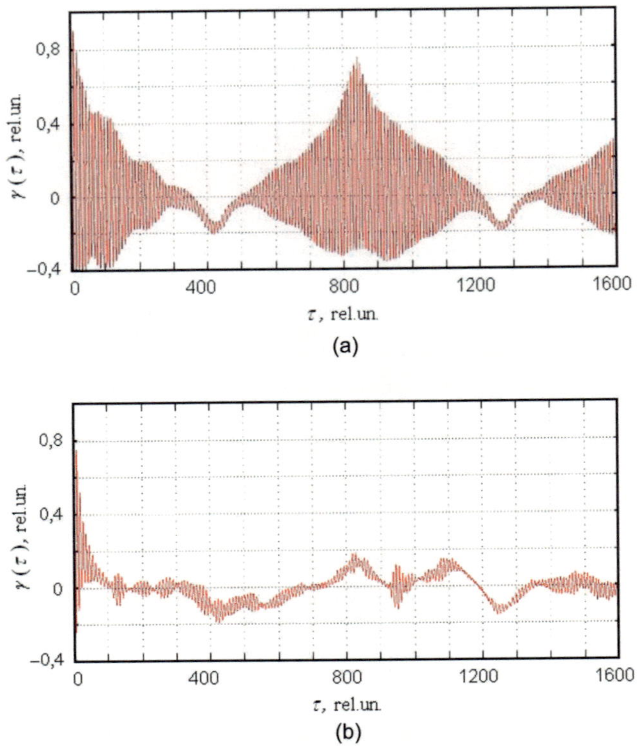

Figure 9.
Auto-correlation function of (a) train of pulses of the kind as shown in **Figure 5** *and (b) quasi-chaotic signal of* **Figure 8** *computed over the time interval* $\tau = 1000 - 9000$ *and* $\tau = 2000 - 20000$*, respectively [23].*

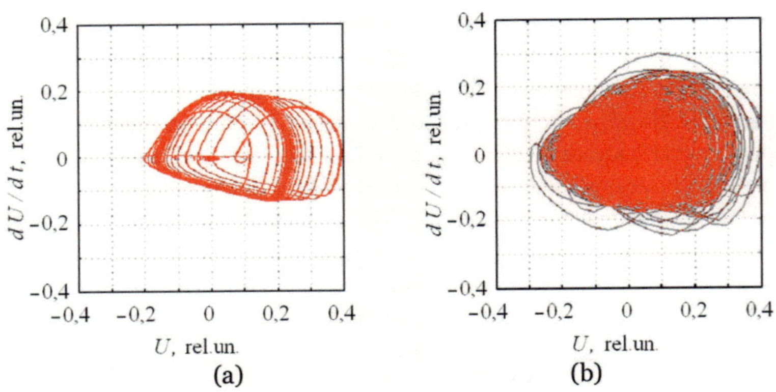

Figure 10.
Poincare sections of (a) pulsed and (b) quasi-chaotic signals processed in **Figure 9** *[23].*

One of the oscillators of such a kind was subject to the frequency-domain analysis for the comparison with time-domain simulations [24]. The oscillator had a single active block, though of slightly different design from those above, and a stub of length S. The analysis followed the ideas of the Kurokawa approach when applied to an open radiating circuit.

The frequency-domain analysis in the form adjusted for the radiating circuits was found to be capable of predicting small-signal oscillation spectra in either single- or multi-frequency cases [24]. At the same time, the approach, naturally, failed when strong nonlinear oscillations of more complicated character have been developed.

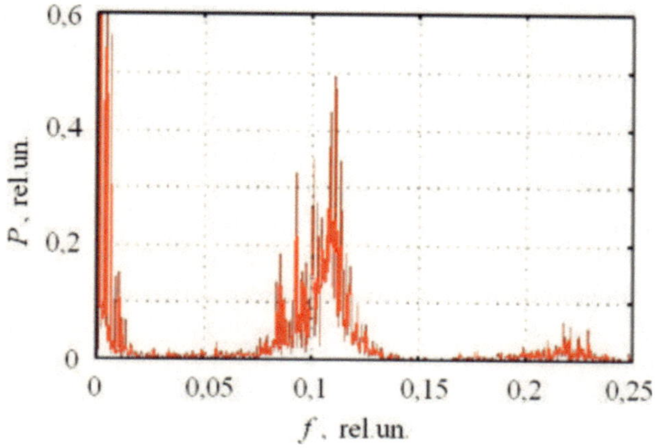

Figure 11.
Frequency spectrum of the waveform presented in **Figure 8** *[23].*

8. Resonant tunneling diode and laser diode circuit emitting trains of correlated EHF and optical pulses

Finally, we consider a time-delay version of an interesting oscillator that uses the laser diode driven by the resonant tunneling diode. The original form of the oscillator was proposed in [32, 33]. The RTD-LD circuit could generate optical (infrared) LD signals (pulses, oscillations, or chaos) when the RTD was excited by the external radio frequency (RF) bias. The authors made up the oscillator operating at the frequencies up to 2 GHz [33] and analyzed it with numerical simulations as a lumped unit.

The LD used in [33] was an optical communication laser operating at around 1550 nm IR radiation wavelength with an average output power of 5 mW. The RTD-LD hybrid circuit was produced with a minimal length of bonding wires b accessible with manual manufacture ($b \sim 1$ mm) so as to minimize the inductance of the system. Using the RTDs of small capacitance, the authors observed self-oscillations at the frequencies of 350–400 MHz, 550–590 MHz, and 1.82–2.17 GHz, depending on the bonding wire length b and other parameters. The authors made numerical simulations of their lumped oscillators and obtained a sufficiently good coincidence with experimental results [33]. The lumped circuit model is perfectly valid at these oscillator sizes and oscillation frequencies.

We considered a distributed version of the circuit that transforms the latter into an open system of the kind as shown in **Figures 1** and **2** of Sections 5 and 6, where the Gunn diode is replaced by a monolithic RTD-LD unit and the remote resonator block A_0 is used as a resonant antenna [25]. At the non-zero length D of the microstrip section connecting the RTD-LD unit and the resonant antenna, the circuit operates as a time-delay oscillator in a self-excitation mode when the RTD bias voltage falls in the NDR domain of the RTD unit.

Under the oscillation conditions, when using appropriate values of circuit parameters and, particularly, choosing the RTD operating at sufficiently high frequencies, e.g., up to 1 THz [34, 35], one can make the circuit to generate a train of short EHF pulses U_A, which are emitted into an open microstrip section and, eventually, radiated into the free space. At the same time, if the LD is also a high-speed device [36], a similar train of optical pulses would be emitted, which are correlated in a perfectly synchronous manner with original EHF pulses.

The trains of both the EHF and optical pulses can be rapidly turned on and off by the external bias digital signal applied to the RTD so as to produce relatively short pieces of trains of EHF pulses. This property, along with a possibility of synchronous generation of both the EHF and optical pulses, can make the dual EHF-optical oscillators of this kind attractive for potential applications in various forms of close range radar systems. The optical power output of these systems can be increased when making reasonably large chip-on-board LD arrays of, e.g., 20 elements or so in a way similar to conventional high-power LED arrays.

There is an optimal range of the length of microstrip section D when the oscillator generates both the optical and EHF pulses being of a good shape, at the maximum EHF frequency, and the maximum light intensity. For example, at the device frequency $f_G = f_A = 7.6$ GHz, the entire range where the effect exists is $D = 1 - 20$ mm and the optimal D is about $D = 5$ mm ($d = 5$ at the unit length $a = 1$ mm).

In addition to the basic form of dual EHF-optical pulse oscillations presented above, there is another, a more interesting mode of operation. When having a sufficiently large excess in the operation speed of the RTD-LD unit as compared to the duration of separate pulses, each optical pulse can also be modulated in power by the RTD EHF current. This makes both the EHF and optical pulses carrying the same EHF signal, **Figure 12**.

Figure 12 shows an example of the EHF pulse radiated by the RTD-LD circuit and the relevant EHF-modulated optical pulse. Pulse oscillations have been computed when both the RTD-LD and the resonator circuit intrinsic frequencies are $f_G = f_A = 76$ GHz assuming $\omega_C = \omega_{C_A} = 1$, $\omega_L = \omega_{L_A} = 1$, and $d = 1$. In this case, the LD electron lifetime parameter is chosen to be $t_n = 10$ ps so that the LD optical output could follow more frequent oscillations of the RTD circuit. These values assume sufficiently small capacitance and inductance of the devices. At the given parameters, they should be about $C = C_A = 55$ fF and $L = L_A = 80$ pH. Despite being extremely small, these values become accessible with modern RTD technology [34, 35]. When using, instead, the parameters $\omega_C = \omega_{C_A} = 10$ and $\omega_L = \omega_{L_A} = 0.1$, we get $C = C_A = 5.5$ fF and $L = L_A = 0.8$ nH that makes, possibly, a more realistic system. In this case, we can produce a single oscillation within a pulse and a short burst of the EHF oscillations at the pulse length $t_P = 25$ ps and $t_P = 50$ ps, respectively [25].

Detection of the EHF modulation of optical (IR) pulses is yet another issue. Direct detection of the EHF modulation of optical pulses is expected to become possible with an RF optical heterodyne photo-detector proposed in [37].

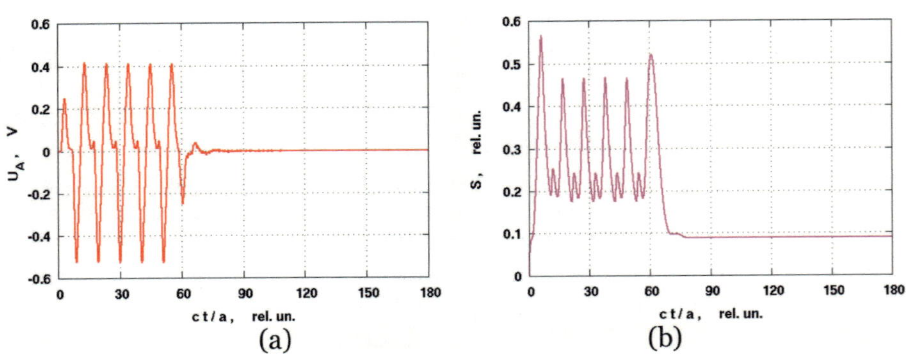

Figure 12.
Excitation of (a) the EHF pulse and (b) the EHF-modulated optical pulse when $G_0 = 0.5$, $d = 1$, $f_G = f_A = 76$ GHz, and $t_P = 0.2$ ns [25].

9. Conclusions

We presented time-domain simulations of distributed time-delay oscillators, which show complicated nonlinear dynamics of electromagnetic oscillations generated by these systems.

We considered the models that consist of discrete lumped circuits of active and passive devices and distributed sections of microstrip lines connecting the lumped circuits. The active devices were presented by the Gunn diodes and the resonant tunneling diodes operating in the EHF frequency band. Parallel and series connections of microstrip sections with active devices have been simulated. Other structures like a 2D microwave cavity with a wall covered with active devices and a 1D open resonator made by a similar wall and a thin dielectric mirror have also been investigated.

Simulation models were developed, which rely on the method that reduces the problems for the wave equations in structures with active devices to the problems with time-delay equations of difference-differential kind. A Dormand-Prince method of the (5,3) order of accuracy for ordinary differential equations was applied and extended for solving time-delay equations arising in simulation.

Time-domain simulations revealed a diversity of dynamical effects in time-delay oscillators being considered. A possibility of chaotic or quasi-chaotic oscillations was observed in a 2D cavity with active devices and in some cases of parallel connection of microstrip sections with Gunn diode circuits.

Self-excitation of trains of the EHF pulses in microstrip structures with Gunn diodes and remote resonators have been discovered. A similar kind of trains of either the baseband or the EHF pulses were also observed in a more simplified model of a 1D cavity with an active layer and a dielectric mirror. Bistability in the generation of either the continuous waves or the trains of the EHF pulses in the Gunn diode systems with remote resonator was discovered and hysteresis in switching between these generation modes was observed.

A dual kind of the EHF-pulse and the EHF-modulated optical pulse generator using an RTD-LD time-delay oscillator has been proposed and investigated.

The approach based on the splitting of oscillator systems on discrete parts of active circuits and distributed parts of propagation sections makes it possible time-domain simulations of complex oscillatory structures. The approach has a major advantage over exact modeling with common engineering software. It allows one to address the problems of time-domain simulation of distributed systems with limited computational resources. This makes the analysis of this kind of systems feasible that, otherwise, would be out of reach with conventional simulation tools.

Acknowledgements

This work was funded in part by NATO research projects NUKR.SFPP 985465 (G5465) and NUKR.SFPP 985395 (G5395) within 37 frames of the Science for Peace and Security (SPS) program.

Author details

Vladimir Yurchenko* and Lidiya Yurchenko
O. Ya. Usikov Institute for Radiophysics and Electronics, National Academy of Sciences of Ukraine, Kharkiv, Ukraine

*Address all correspondence to: v.yurchenko.nuim@gmail.com

References

[1] Lukin K. Noise radar technology. Telecommunications and Radio Engineering. 2001;**55**(12):8-16. DOI: 10.1615/TelecomRadEng.v55.i12

[2] Lukin K. Millimeter-wave band noise radar. Telecommunications and Radio Engineering. 2009;**68**(14):1229-1255. DOI: 10.1615/TelecomRadEng.v68.i14.20

[3] Golio M, editor. The RF and Microwave Handbook. Boca Raton: CRC Press LLC; 2000. 1376 p. ISBN 9781420036763

[4] Grebennikov A. RF and Microwave Transistor Oscillator Design. Chichester: Wiley; 2007. 441 p. DOI: 10.1002/9780470512098

[5] Kurokawa K. Noise in synchronized oscillators. IEEE Transactions on Microwave Theory and Techniques. 1968;**16**:234-240. DOI: 10.1109/TMTT.1968.1126656

[6] Kurokawa K. The single-cavity multiple-device oscillator. IEEE Transactions on Microwave Theory and Techniques. 1971;**19**:793-801. DOI: 10.1109/TMTT.1971.1127642

[7] Vendelin G, Pavio A, Rohde U. Microwave Circuit Design Using Linear and Nonlinear Techniques. Chichester: Wiley; 2005. 1058 p. DOI: 10.1002/0471715832

[8] Gonzalez-Posadas V, Jimenez-Martin J, Parra-Cerrada A, Segovia-Vargas D, Garcia-Munoz L. Oscillator accurate linear analysis and design. Classic linear methods review and comments. Progress in Electromagnetics Research. 2011;**118**:89-116. DOI: 10.2528/PIER11041403

[9] Erturk V, Rojas R, Roblin P. Hybrid analysis/design method for active integrated antennas. IEE Proceedings Microwave Antennas Propagation. 1999;**146**(2):131-137. DOI: 10.1049/ip-map:19990208

[10] Rao S. Time Domain Electromagnetics. San Diego: Academic Press; 1999. 372 p. DOI: 10.1016/B978-0-12-580190-4.X5000-9

[11] Aksoy S, Tretyakov O. Evolution equations for analytical study of digital signals in waveguides. Journal of Electromagnetic Waves and Applications. 2003;**17**(12):1665-1682. DOI: 10.1163/156939303322760209

[12] Sirenko Y, Strom S, Yashina N. Modeling and Analysis of Transient Processes in Open Resonant Structures. New York: Springer; 2007. 353 p. DOI: 10.1007/0-387-32577-8

[13] Asakawa K, Itagaki Y, Shi-Ya H, Saito M, Suhara M. Time-domain analysis of large-signal-based nonlinear models for a resonant tunneling diode with an integrated antenna. IEICE Trans. on Electronics. 2012;**E95-C**(8):1376-1384. DOI: 10.1587/transele.E95.C.1376

[14] Coillet A, Henriet R, Salzenstein P, Huy KP, Larger L, Chembo YK. Time-domain dynamics and stability analysis of optoelectronic oscillators based on whispering-gallery mode resonators. IEEE Journal of Selected Topics in Quantum Electronics. 2013;**19**(5):6000112. DOI: 10.1109/JSTQE.2013.2252152

[15] Anzill W, Russer P. A general method to simulate noise in oscillators based on frequency domain techniques. IEEE Trans. on Microwave Theory and Techniques. 1993;**41**(12):2256-2263. DOI: 10.1109/22.260715

[16] Atay F, editor. Complex Time-Delay Systems. Berlin: Springer-Verlag; 2010. 322 p. DOI: 10.1007/978-3-642-02329-3

[17] Yurchenko L, Yurchenko V. Noise generation in a cavity resonator with a wall of solid-state power-combining array. In: Proceedings of the 11th International Conference on Microwaves and Radar (MIKON-96); 27-30 May 1996; Poland. Vol. 2. Warsaw: IEEE; 1996. pp. 454-458

[18] Yurchenko L, Yurchenko V. Analysis of the dynamical chaos in a cavity with an array of active devices. In: Proceedings of the 12th International Conference on Microwaves and Radar (MIKON-98); 20–22 May 1998; Poland. Vol. 3. Krakow: IEEE; 1998. pp. 723-727

[19] Yurchenko L, Yurchenko V. Self-generation of ultra-short pulses in a cavity with a dielectric mirror excited by an array of active THz devices. In: Proceedings of the 8th International Conference on Terahertz Electronics. 28–29 September 2000, Darmstadt, Germany; 2000. pp. 49-52

[20] Yurchenko V, Yurchenko L. Bistability and hysteresis in the emergence of pulses in microstrip Gunn-diode circuits. AIP Advances. 2014;**4**:127126. DOI: 10.1063/1.4904226

[21] Yurchenko L, Yurchenko V. Time-domain simulation of power combining by a parallel connection of strip lines with Gunn diodes. Telecommunications and Radio Engineering. 2014;**73**(5): 375-390. DOI: 10.1615/TelecomRadEng. v73.i5.10

[22] Yurchenko V, Yurchenko L. Time-domain simulation of power combining in a chain of THz Gunn diodes in a transmission line. International Journal of Infrared and Millimeter Waves. 2004;**25**(1):43-54. DOI: 10.1023/B: IJIM.0000012761.27947.01

[23] Yurchenko L, Yurchenko V. Time-domain simulation of short-pulse oscillations in a Gunn diode system with time-delay microstrip coupling. Applied Radio Electronics. 2013;**12**(1):45-50. Print ISSN 1727-1290

[24] Yurchenko V, Yurchenko L. Self-excitation of a chain of Gunn diodes connected by transmission lines. In: Proceedings of the 2nd URSI-Turkey'2004 Symposium; 8–10 September 2004; Turkey, Ankara: URSI; 2004. pp. 460-462

[25] Yurchenko V, Yurchenko L, Ciydem M. Pulse-mode simulations of RTD-LD circuits for visible light communication. Applied Radio Electronics. 2018;**17**(1-2): 66-71. Print ISSN 1727–1290

[26] Hairer E, Wanner G. Solving Ordinary Differential Equations II: Stiff and Differential-Algebraic Problems. 2nd ed. Berlin: Springer-Verlag; 1996. 609 p. DOI: 10.1007/978-3-662-09947-6

[27] SmartSpice. [Internet]. 2018. Available from: https://www.silvaco.com/products/analog_mixed_signal/smartspice_46.pdf [Accessed: 2018-09-24]

[28] Yurchenko V, Ciydem M, Gradziel M, Yurchenko L. MM-wave dielectric parameters of magnesium fluoride glass wafers. Progress in Electromagnetics Research. 2017;**62**:89-98. DOI: 10.2528/PIERM17081805

[29] Zou W, Senthilkumar D, Duan J, Kurths J. Emergence of amplitude and oscillation death in identical coupled oscillators. Physical Review E. 2014;**90**: 032906. DOI: 10.1103/PhysRevE.90.032906

[30] Masoller C, Sukow D, Gavrielides A, Sciamanna M. Bifurcation to square-wave switching in orthogonally delay-coupled semiconductor lasers: Theory and experiment. Physical Review A. 2011;**84**:023838. DOI: 10.1103/PhysRevA.84.023838

[31] DeLisio M, York R. Quasi-optical and spatial power combining. IEEE Transactions on Microwave Theory and

Techniques. 2002;**50**(3):929-936. DOI: 10.1109/22.989975

[32] Slight T, Romeira B, Wang L, Figueiredo J, Wasige E, Ironside C. A Lienard oscillator resonant tunnelling diode-laser diode hybrid integrated circuit: Model and experiment. IEEE Journal Quantum Electronics. 2008; **44**(12):1158-1163. DOI: 10.1109/JQE.2008.2000924

[33] Romeira B, Figueiredo J, Slight T, Wang L, Wasige E, Ironside C, et al. Synchronisation and chaos in a laser diode driven by a resonant tunnelling diode. IET Optoelectronics. 2008;**2**: 211-215. DOI: 10.1049/iet-opt:20080024

[34] Suzuki S, Asada M, Teranishi A, Sugiyama H, Yokoyama H. Fundamental oscillation of resonant tunneling diodes above 1 THz at room temperature. Applied Physics Letters. 2010;**97**(24):242102. DOI: 10.1063/1.3525834

[35] Feiginov M, Sydlo C, Cojocari O, Meissner P. Resonant-tunnelling-diode oscillators operating at frequencies above 1.1 THz. Applied Physics Letters. 2011;**99**:233506. DOI: 10.1063/1.3667191

[36] Tsonev D, Videv S, Haas H. Towards a 100 Gb/s visible light wireless access network. Optics Express. 2015;**23**(2):1627-1637. DOI: 10.1364/OE.23.001627

[37] Li Q, Sun K, Li K, Yu Q, Runge P, Ebert W, et al. High-power evanescently coupled waveguide MUTC photodiode with > 105-GHz bandwidth. Journal of Lightwave Technology. 2017; **35**(21):4752-4757. DOI: 10.1109/JLT.2017.2759210

Chapter 5

Oscillator Dampers in Civil Structures

Yonggang Tan

Abstract

Many kinds of oscillators, springs, and damping system compose vibration reduction system in civil structures. Since the invention of the tuned mass damper (TMD) device a century ago, it has become a very important technology in structural control. TMDs can effectively suppress the response of civil structures under harmonic or wind excitations. To improve the damping capacity of TMDs in reducing the vibration of structures under seismic loads, a large mass ratio should be used, but TMDs are still ineffective in suppressing the seismic peak response of high-rise buildings. The inerter-based dynamic vibration absorbers (IDVA), including tuned inerter dampers (TID) and tuned mass-damper-inerter (TMDI), have been investigated in recent years. The advantage of using a TID and TMDI comes from the adoption of gearing in the inerter, which equivalently amplifies the mass. The mass ratio of an inerter is very high; hence, its mechanical properties and reliability are vital. A novel damper device, accelerated oscillator damper (AOD), has been proposed recently. Gear transmission systems are used to generate an amplified kinetic energy of the oscillator to reduce the oscillations of the structures. The AOD system is superior to the traditional TMD system in short time loading intervals or under the maximum seismic loads.

Keywords: TMD, AOD, accelerated oscillator damper, inerter-based dynamic vibration absorber, tuned inerter damper, tuned mass-damper-inerter

1. Introduction

Oscillations are frequent motions between two extreme amounts or more different states. The term vibration is formally used to describe mechanical oscillation. Oscillations occur in dynamic systems in almost every area of science and engineering, for instance, structural vibration induced by earthquake, water waves and wind, vibrating strings in musical instruments, and beating of the human heart. Some of the oscillations are harmful and even destructive; so many devices and facilities have been used to reduce the undesirable vibrations of structures. Familiar measures of oscillation reduction are using springs, rubber, viscous dampers, and so on.

The oscillator damper is a kind of device designed for suppressing oscillations in structures and machines. The principle of oscillator dampers is to use mass oscillators generating opposite forces acting on the structures to resist their oscillations. In 1909, Frahm [1] proposed the first spring and mass oscillator damper system for suppressing the mechanical vibration induced by harmonic forces. In 1928, Den Hartog and Ormondroyd [2] added a certain damping to the Frahm oscillator damper model, which is the prototype of tuned mass dampers (TMD). Den Hartog [3] provided a detailed description and design formulas for TMD. In the early stage, the application of TMD mainly focused on the vibration problem of mechanical systems. Since 1971, the TMD systems have been widely used in civil structures such as super high-rise buildings, tower buildings, towers and decks of cable-stayed bridges, and suspension bridges. These applications show that TMDs can effectively reduce structural vibrations induced by wind. However, TMDs may encounter some disadvantages in the application of earthquake: large stroke and detuning problems and large seismic forces within short time intervals. To solve the detuning problem, Xu and Igusa [4] proposed an improved vibration absorber, called multiple tuned mass damper (MTMD), which is less sensitive to frequency change. Parametric optimization and control efficiency of MTMD systems were studied, and various design theories were established in the 1990s [5–7]. A tuned liquid damper (TLD) is a type of oscillator damper with the solid mass being replaced by liquid. The TLD consists of rigid rectangular tanks partially filled with liquid. Its damping effect comes from liquid sloshing forces or moments, which can change the dynamic properties of the structure and reduce the dynamic response of the system subjected to external excitation [8]. The application of TLD for the vibration control of civil structures was studied in the late 1980s Fujino et al. [8]. The success of the TLD system in reducing wind-induced structural vibrations has been well established with the support of numerous analytical and experimental studies.

The tuned inerter damper (TID) is a new form of TMD with the mass being replaced by an inerter. The inerter is a two-terminal mechanical device developing a resisting force proportional to the relative acceleration of its terminals [9]. A simple approach to constitute an inerter is to have a rod sliding in linear bearings, which drives a flywheel via a rack, pinion, and gear. The advantage of using a TID comes from the adoption of gearing in the inerter, which equivalently amplifies the mass [10]. TID can be applied to reduce vibrations in civil structures subjected to base excitation.

In 2015, Giaralis and Marian [11] proposed the generalization of the classical TMD, which introduced an inerter: the tuned mass-damper-inerter (TMDI). With similar working principle as the TID, the TMDI can be used to reduce structural oscillations excited by stochastic loadings [12].

Recently, a novel effective device, accelerated oscillator damper (AOD), was proposed to suppress the vibrations of civil structures under seismic loads [13]. AOD includes oscillator mass, transmission, spring, and viscous damper. The kinetic energy of the appended oscillator is proportional to the square of its velocity. The rack and gear transmission system enlarged the speed of the oscillator mass. As a result, the kinetic energy of the appended oscillator is amplified, leading to the kinetic energy of the structures being reduced for the principle of energy conservation. It was found that the AOD system is superior to the traditional TMD system in short time loading intervals or under the maximum seismic loads.

2. Tuned mass damper

The concept of TMDs was proposed by Frahm and applied for the patent of the United States in 1909. After more than 100 years of development, it has become the most popular type of damper. In the early stage, the application of TMD mainly focused on the vibration problem of mechanical systems. Since 1971, the TMD systems have been widely used in civil structures such as super high-rise buildings, tower buildings, towers and decks of cable-stayed bridges, and suspension bridges.

2.1 Equations of motion of the TMD system

According to **Figure 1**, the equations of motion of a single degree of freedom (SDOF) structure-TMD system are given as [18]

$$M\ddot{X}(t) + KX(t) - c\{\dot{x}(t) - \dot{X}(t)\} - k\{x(t) - X(t)\} = P(t) \tag{1}$$

$$m\ddot{x}(t) + c\{\dot{x}(t) - \dot{X}(t)\} + k\{x(t) - X(t)\} = p(t) \tag{2}$$

where M is the primary mass, m is the secondary mass, K is the primary spring stiffness, k is the secondary spring stiffness, c is the secondary damping, $P(t)$ is the force acting on primary mass, and $p(t)$ is the force acting on damper mass.

For further discussion, other symbols are introduced as follows:

ω is the frequency of a harmonic excitation; Ω is the natural frequency of primary mass, $\Omega = \sqrt{K/M}$; ω_a is the natural frequency of secondary mass, $\omega_a = \sqrt{k/m}$; μ is the damper mass to primary mass ratio, $\mu = m/M$; g_1 is the ratio of excitation frequency to primary mass natural frequency, $g_1 = \omega/\Omega$; f is the frequency ratio, $f = \omega_a/\Omega$; ζ_d is the damping ratio of TMD; and ζ is the damping ratio of primary mass.

Den Hartog [3] studied closed form expressions of optimal damper parameters f and ζ_d, which minimize the steady-state response of the primary mass subjected to a harmonic excitation. Optimal damper parameters can be calculated by the following equations:

$$f_{opt} = \frac{1}{1+\mu} \tag{3}$$

$$\zeta_{dopt} = \sqrt{\frac{3\mu}{8(1+\mu)}} \tag{4}$$

When the structure is subjected to a harmonic base excitation, the optimal damper parameters can be expressed as

$$f_{opt} = \frac{1}{1+\mu}\sqrt{\frac{2-\mu}{2}} \tag{5}$$

$$\zeta_{dopt} = \sqrt{\frac{3\mu}{8(1+\mu)}}\sqrt{\frac{2-\mu}{2}} \tag{6}$$

2.2 Mechanical performance of TMDs

TMD is a passive energy absorbing device attached to a vibrating primary structure to reduce undesirable vibrations [15]. It is found that if a secondary system is implemented on a primary structure and its natural frequency is tuned to be very close to the dominant mode of the primary structure, a large reduction in the dynamic responses of the primary structure can be achieved [14, 16]. Tuned mass dampers are effective in reducing the response of structures due to harmonic [17] or wind [18] excitations. A steady-state harmonic analysis of the effect of detuning with varying excitation frequencies was investigated by Rana and Soong. It was found that if the TMD parameters shift away from their optimal values, the response control is expected to degrade.

Although the basic design concept of TMD is very simple, the parameters (mass, stiffness, and damping) of the TMD system must be obtained by optimal design

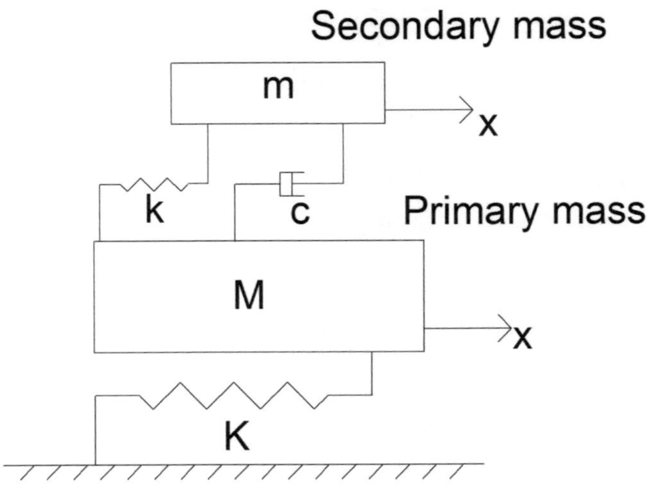

Figure 1.
A schematic representation of the TMD system.

procedures to attain better control performance. Therefore, the determination of optimal design parameters of TMD to enhance the control effectiveness has become very crucial [16].

TMDs have many advantages, such as simplicity, effectiveness, and low cost [19]. However, single tuned mass damper (STMD) is sensitive to the frequency ratio between the TMD and the structure, and it is also sensitive to the damping ratio of TMD. As a result, the use of more than one TMD with different parameters has been proposed by Xu and Igusa [4] to improve the effectiveness and robustness.

2.3 Multiple tuned mass dampers

In order to further improve the shortcomings of the robustness and effectiveness of TMD, Igusa and Xu proposed a multiple tuned mass damper with multiple different dynamic characteristics and a linear distribution of frequency, namely MTMD. MTMD and TMD work basically the same way, except that MTMD is composed of multiple TMDs. They work together under external load excitation to achieve the best vibration absorption.

As shown in **Figure 2**, m_s is the mass of the primary structure and k_s and c_s are the stiffness and damping of the primary structure respectively; MTMD is mainly composed of n TMDs (expressed by subscripts from 1 to n), and the corresponding mass, stiffness, and damping parameters (m, k, and c) of each TMD may be different (denoted by subscripts from 1 to n). In fact, these parameters are usually different for better vibration control, but the frequency of each TMD is centered on the frequency of the main control mode. When the main structure is excited by external loads, the mass m that is out of phase with the main structure is applied to the main structure with a force opposite to the direction of motion, thereby achieving the purpose of damping.

The advantages of MTMDs are as follows: (1) compared with TMDs, MTMDs are more suitable for controlling structural vibration of frequency changes, because TMD is a single frequency, and MTMD is composed of multiple TMDs of different frequencies, which can adapt to a wider bandwidth, that is, more robust; (2) MTMDs are more achievable than TMDs, because the weight of the mass of a single TMD is generally 1–4% of the mass of the main structure (a large concentrated load), which may cause local damage to the building where TMD is installed, but

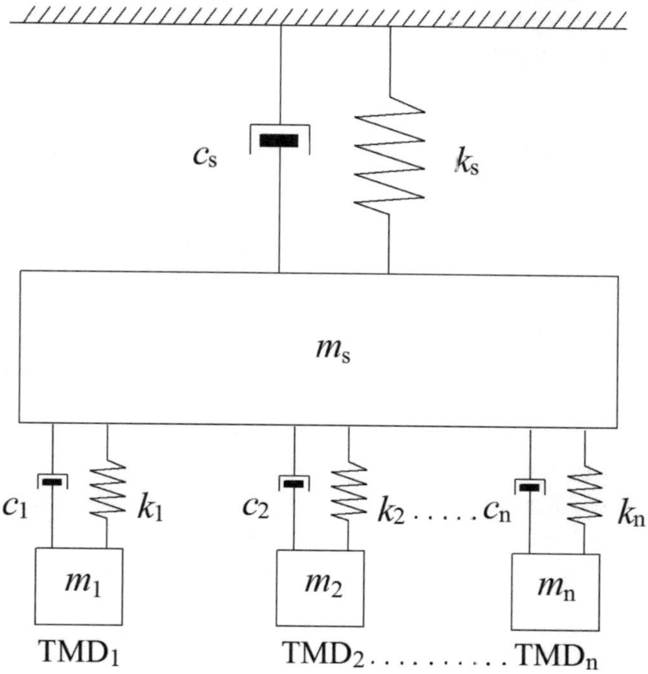

Figure 2.
A schematic representation of the MTMD system.

MTMD consists of multiple TMDs, effectively dispersing the weight of the mass, is small in size and achievable; (3) simple installation, convenient maintenance, and low cost.

However, the parameter optimization of MTMD is more difficult. How many TMDs should be used in a structure? How to choose the parameters of each TMD? Where each TMD should be installed? All these questions should be solved properly. Moreover, the selection of these parameters will be affected by the site conditions, so the problem of parameter optimization remains to be further studied.

TMDs and MTMDs have been installed in high-rise buildings or pedestrian bridges to reduce wind-induced vibrations. Typical examples include: the John Hancock Tower in Boston, the Citicorp Center Office Building in New York City, the Terrace on the Park Building in New York City, and the Taipei 101 Tower in Taiwan [16].

2.4 Tuned liquid damper

A tuned liquid damper (TLD) is a type of TMD where the mass is replaced by a liquid. The TLD consists of rigid rectangular tanks partially filled with liquid. Its damping effect comes from liquid sloshing forces or moments, which can change the dynamic properties of the structure and reduce the dynamic response of the system subjected to external excitation. By changing the basic sloshing frequency of the TLD close to the natural frequency of the structure, the inertia forces could act on the opposite direction to the external excitation force, which reduces the response of the structure with a TLD. The natural frequency of TLD can be controlled by adjusting the depth of liquid and dimension of container.

Since TLD has many advantages over other conventional dampers, it attracted a lot of attention to reduce vibrations in many applications. It requires little maintenance and is easy to install in civil structures [20]. The response of a typical SDOF structure is reduced by approximately 30% if a TLD has a depth ratio of 0.15 and a mass ratio of 4%. The application of TLD has been used as passive control devices to control the vibrations of civil structures under dynamic loads induced by wind and earthquake.

2.5 Active tuned mass damper and semi-active tuned mass damper

With the development of computer technology and modern control theory, structural control technology extends from passive control to active control and semi-active control. Based on the passive controlled TMD, active tuned mass damper (ATMD) was introduced with an active controller using an external source of power to generate additional forces on structures, and optimization procedures were proposed to compute the required control forces. Therefore, ATMD is effective in suppressing seismic response and more robust to mistuning with appropriate usage of feedback. Since the 1980s, the ATMD control systems for civil engineering structures have attracted considerable attention [21].

The ATMD control system is composed of three sub-systems, namely sensor, control decision maker, and ATMD device. An active control mechanism is included between the SDOF corresponding to the building model and the damper mass [22].

The motion equations of the MDOF building are expressed as

$$[M]\{\ddot{x}\} + [C]\{\dot{x}\} + [K]\{x\} = \{F\} + \{B_0\}u \tag{7}$$

where $[M]_{n\times n}$, $[C]_{n\times n}$, and $[K]_{n\times n}$ are the mass, damping, and stiffness matrices of the structure, respectively; $\{x\}_{n\times 1}$ is the vector that contains the displacement degrees of freedom; $\{F\}_{n\times 1}$ is the vector that contains the external excitation forces; $\{B_0\}_{n\times 1}$ is the vector that describes the location of the control; and u is the scalar control; if $u = 0$, there is no active control input to the structure.

The semi-active tuned mass damper (semi-ATMD) is a device with time varying controllable damping replacing the active controller of ATMD. Compared with classical active dampers, the semi-ATMD requires a small amount of active force or energy to change the valve of damping, but does not dissipate the total energy of the structures directly. In a sense, the semi-ATMD can be more likely treated as a passive device rather than pure active tuned mass damper.

3. Inerter-based dynamic vibration absorber

3.1 Tuned inerter damper

The tuned inerter damper (TID) is a new form of TMD with the mass being replaced by an inerter. The inerter is, as is illustrated in **Figure 3**, a two-terminal mechanical device developing a resisting force proportional to the relative acceleration of its terminals. A simple approach to constitute an inerter is to have a rod sliding in linear bearings, which drives a flywheel via a rack, pinion, and gear.

TIDs, as are shown in **Figure 4**, offer a promising alternative to TMDs due to the fact that inerters, which produce a force proportional to the relative acceleration of their terminals, are geared and can produce a far larger apparent mass than the actual device mass. Therefore, the modal damping ratio obtained via TID can be higher than that achieved by a traditional viscous damper or TMD. A commercially available inerter, the Penske 8760H, has an apparent mass (inertance) to device mass ratio of 38 (higher mass ratios have been reported such as 200), whereas TMD has a general mass ratio of 10%.

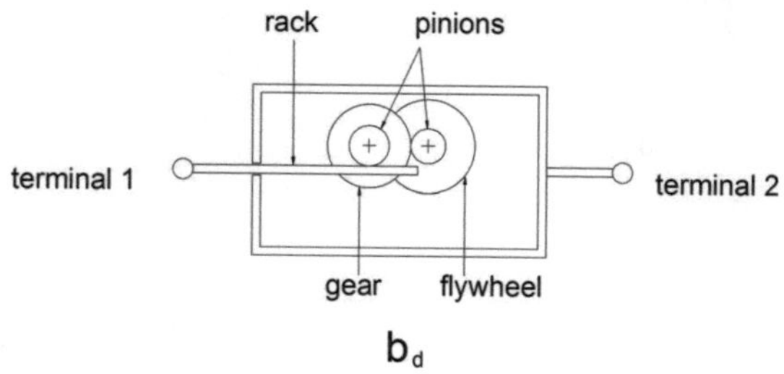

Figure 3.
Schematic representation of the two-terminal flywheel device.

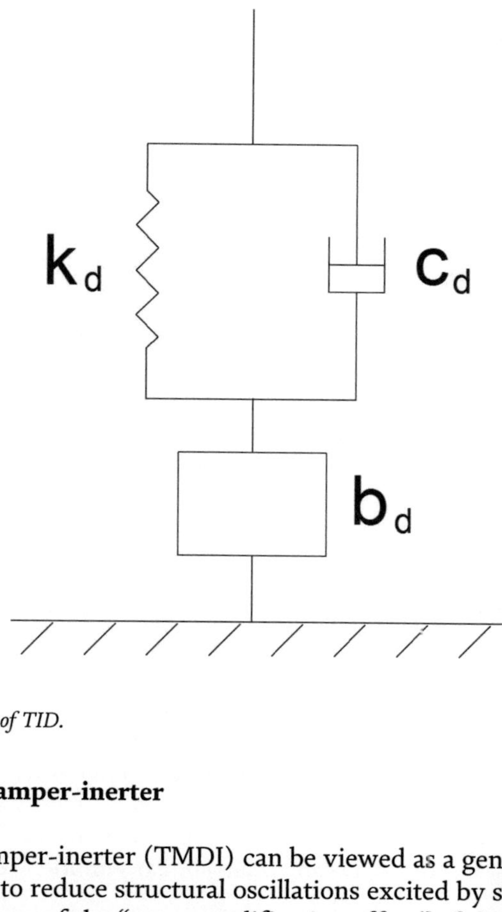

Figure 4.
Schematic demonstration of TID.

3.2. Tuned mass damper-inerter

Tuned mass damper-inerter (TMDI) can be viewed as a generalization of the conventional TMD to reduce structural oscillations excited by stochastic loadings. TMDI takes advantage of the "mass amplification effect" of the inerter to achieve enhanced performance compared to the classical TMD.

Figure 5 shows the SDOF primary structure incorporating the TMDI system configuration. The motion equations of the linear dynamical system shown in **Figure 5** can be expressed as

$$\begin{bmatrix} m_{TMD}+b & 0 \\ 0 & m_1 \end{bmatrix} \begin{Bmatrix} \ddot{x}_{TMD} \\ \ddot{x}_1 \end{Bmatrix} + \begin{bmatrix} m_{TMD}+b & 0 \\ 0 & m_1 \end{bmatrix} \begin{Bmatrix} \dot{x}_{TMD} \\ \dot{x}_1 \end{Bmatrix} + \begin{bmatrix} m_{TMD}+b & 0 \\ 0 & m_1 \end{bmatrix} \begin{Bmatrix} \dot{x}_{TMD} \\ \dot{x}_1 \end{Bmatrix} = -\begin{Bmatrix} \dot{x}_{TMD} \\ \dot{x}_1 \end{Bmatrix} a_g \tag{8}$$

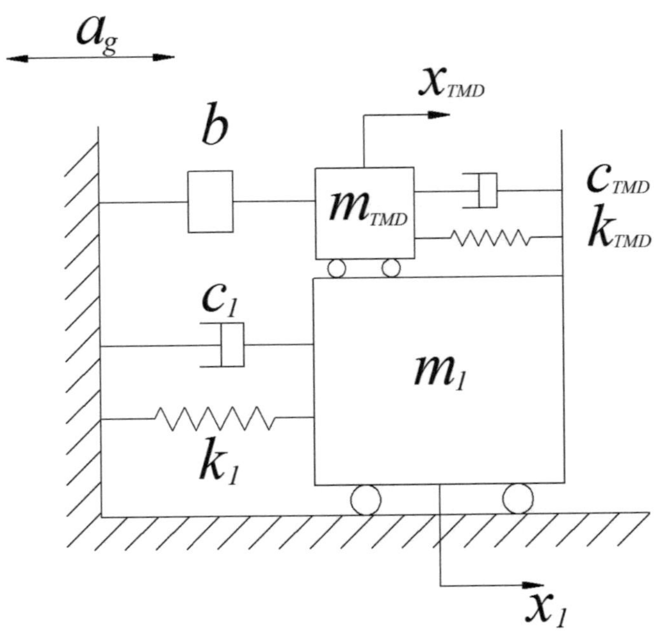

Figure 5.
SDOF primary structure-TMDI system.

It was proved that the optimum designed TMDI system is more effective than the conventional TMD in reducing the displacement variance of white noise excited undamped SDOF primary structures.

The main application of TMDI is used as passive control devices to suppress the vibrations in civil engineering structures under dynamic loads, such as vehicles, wind, rain, earthquake, and so on.

4. Accelerated oscillator damper

A novel damper device, accelerated oscillator damper (AOD), has been proposed recently [13]. The AOD system includes oscillator mass, transmission, spring, and viscous damper. As is illustrated in **Figure 6**, the oscillatory motion of the primary structure is transferred by a geared transmission to enlarge the velocity of the secondary oscillator mass. Rather than driving of a fly-wheel inerter, AOD amplifies and transfers the motion of primary structure to another larger secondary oscillator mass than inerter mass. Therefore, AOD can obtain similar vibration reduction

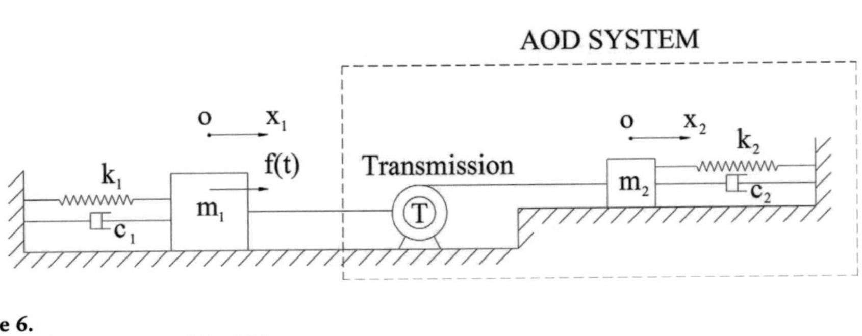

Figure 6.
Schematic demonstration of the AOD system.

effectiveness to TID and TMDI, but the transmission system does not need a very high transmission ratio, which is easier to achieve in engineering practice.

The kinetic energy of the appended oscillator is proportional to the square of its velocity, and rack and gear transmission system enlarged the speed of the oscillator mass. As a result, the kinetic energy of the oscillator mass is also amplified, leading to the kinetic energy of the primary structure being reduced for the principle of energy conservation.

Motion equations of the accelerated oscillator damper system were established by Tan [13] as

$$\ddot{x}_1 + \frac{c_1 + r^2 c_2}{m_1 + r^2 m_2}\dot{x}_1 + \frac{k_1 + r^2 k_2}{m_1 + r^2 m_2}x_1 = \frac{f(t)}{m_1 + r^2 m_2} \tag{9}$$

where x_1 and $\dot{x}_1$ are the displacement and velocity of the primary structure; m_1 and m_2 are the primary and oscillator mass; k_1 is the primary structure stiffness; c_1 is the primary structure damping constant; $f(t)$ is the external force; r is the transmission ratio; x_2 and $\dot{x}_2$ are the displacement and velocity of the oscillator; k_2 is the appended secondary structure spring stiffness; and c_2 is the appended secondary structure damping constant.

A multiple accelerated oscillator damper (MAOD) is defined as multiple AOD devices parallelly attached to the primary structure. Both AOD and MAOD systems can be regarded as generalized SDOF systems. They have the same motion equation forms as the conventional SDOF system.

The effect of the mass ratio of AOD or MAOD is similar to that of the TMD systems, but the ratio of transmission plays more important roles in vibration reduction. The AOD or MAOD devices, with the transmission ratio larger than 2, can achieve a remarkable damping effect. The mass ratio of the AOD or MAOD (sum of total oscillator mass) system can be generally selected below 1%.

It was found that AOD and MAOD systems are more effective than conventional TMD systems in short time loading intervals and the maximum seismic loads. Therefore, they can be used to reduce vibrations in civil structures under wind and seismic excitations.

5. Conclusions

Oscillations, induced by vehicles, wind, water waves, earthquake, and other dynamic loadings, are universal type of motions in mechanical structures and civil structures. Varieties of vibration reduction devices have been proposed to reduce the undesirable oscillations in every field. Oscillator dampers, typically using the inertia force of oscillators to suppress the vibrations of primary structures, are often used as TMD, MTMD, TLD, ATMD, semi-ATMD, TID, TMDI, AOD, MAOD, and so on.

TMDs are effective in reducing the response of structures due to harmonic or wind excitations, but detuning makes the response control degrade. MTMDs are more robust than TMDs and adapt to wider bandwidth. TLDs require little maintenance and operating cost and are easy to install in existing building structures. Parametric optimization is significant for TMDs, MTMDs, and TLDs because it determines the damping efficiency and robustness of the dampers.

The inerter-based dynamic vibration absorber includes TID and TMDI. The advantage of using a TID and TMDI comes from the adoption of gearing in the inerter, which equivalently amplifies the mass of dampers.

AOD is a mass damper composed of oscillator mass, transmission, spring, and viscous damper. The oscillatory motion of the primary structure is transmitted by a geared transmission to enlarge the velocity of the secondary oscillator mass. With the kinetic energy of the oscillator being amplified, the vibration of the primary structures is reduced. The transmission ratio shows more effectiveness in vibration reduction than the mass ratio. The AOD system is superior to the traditional TMD system in short time loading intervals or under the maximum seismic loads.

Bridges and high-rise buildings will be subjected to extraordinarily huge loads during natural disasters such as hurricanes and earthquakes, and the security of the structures will face serious challenges. The application of oscillator dampers can reduce the structural damage caused by vibrations to some extent, which prevents the civil structures from destruction during natural disasters.

Acknowledgements

The authors gratefully acknowledge the financial support from the Natural Science Foundation of China, No. 51008047.

Conflict of interest

We declare that we have no conflict of interest.

Author details

Yonggang Tan
Faculty of Infrastructure Engineering, Dalian University of Technology, Dalian, China

*Address all correspondence to: ygtan@dlut.edu.cn

References

[1] Frahm H. Device for damping vibration of bodies. US Patent No. 989-958; 1911

[2] Den Hartog JP, Ormondroyd J. Theory of the dynamic vibration absorber. Journal of Applied Mechanics-Transactions of the ASME. 1928;**50**: 11-22

[3] Den Hartog JP. Mechanical Vibrations. 4th ed. New York: McGraw Hill; 1956

[4] Xu KM, Igusa T. Dynamic characteristics of multiple substructures with closely spaced frequencies. Earthquake Engineering and Structural Dynamics. 1992;**21**(12):1059-1070

[5] Abe M, Fujino Y. Dynamic characterization of multiple tuned mass dampers and some design formulas. Earthquake Engineering and Structural Dynamics. 1994;**23**(8):813-835

[6] Yamaguchi H, Harnpornchai N. Fundamental characteristics of multiple tuned mass dampers for suppressing harmonically forced-oscillations. Earthquake Engineering and Structural Dynamics. 1993;**22**(1):51-62

[7] Kareem A, Kline S. Performance of multiple mass dampers under random loading. Journal of Structural Engineering, ASCE. 1995;**121**(2): 348-361

[8] Fujino Y, Sun L, Pacheco BM, Chaiseri P. Tuned liquid damper (TLD) for suppressing horizontal motion of structures. Journal of Engineering Mechanics-Asce. 1992;**118**(10): 2017-2030

[9] Smith MC. Synthesis of mechanical networks: The inerter. IEEE Transactions on Automatic Control. 2002;**47**(10):1648-1662

[10] Lazar IF, Neild SA, Wagg DJ. Vibration suppression of cables using tuned inerter dampers. Engineering Structures. 2016;**122**:62-71

[11] Giaralis A, Marian L. Use of inerter devices for weight reduction of tuned mass-dampers for seismic protection of multi-storey buildings: The tuned mass-damper-interter (TMDI). In: Proceedings of Active and Passive Smart Structures and Integrated Systems; 2016

[12] Marian L, Giaralis A. Optimal design of a novel tuned mass-damper-inerter (TMDI) passive vibration control configuration for stochastically support-excited structural systems. Probabilistic Engineering Mechanics. 2014;**38**: 156-164

[13] Tan Y, Guan R, Zhang Z. Performance of accelerated oscillator dampers under seismic loading. Advances in Mechanical Engineering. 2018;**10**(4):1-10

[14] Rana R, Soong TT. Parametric study and simplified design of tuned mass dampers. Engineering Structures. 1998; **20**(3):193-204

[15] Sadek F, Mohraz B, Taylor AW, Chung RM. A method of estimating the parameters of tuned mass dampers for seismic applications. Earthquake Engineering and Structural Dynamics. 1997;**26**(6):617-635

[16] Lee CL, Chen YT, Chung LL, Wang YP. Optimal design theories and applications of tuned mass dampers. Engineering Structures. 2006;**28**(1): 43-53

[17] Luft RW. Optimal tuned mass dampers for buildings. Journal of the Structural Division, ASCE. 1979; **105**(12):2766-2772

[18] McNamara RJ. Tuned mass dampers for buildings. Journal of the Structural Division, ASCE. 1977;**103**(9):1785-1798

[19] Li HN, Ni XL. Optimization of non-uniformly distributed multiple tuned mass damper. Journal of Sound and Vibration. 2007;**308**(1–2):80-97

[20] Nguyen TP, Pham DT, Ngo KT. Effectiveness of multi tuned liquid dampers with slat screens for reducing dynamic responses of structures. IOP Conference Series: Earth and Environmental Science. 2018;**143**: 012023

[21] Chang CC, Yang HTY. Control of buildings using active-tuned mass dampers. Journal of Engineering Mechanics-Asce. 1995;**121**(3):355-366

[22] Ankireddi S, Yang HTY. Simple ATMD control methodology for tall buildings subject to wind loads. Journal of Structural Engineering, ASCE. 1996; **122**(1):83-91